T0212515

SpringerBriefs in Mathematical Physics

Volume 14

SpringerBriefs are characterized in general by their size (50–125 pages) and fast production time (2–3 months compared to 6 months for a monograph).

Briefs are available in print but are intended as a primarily electronic publication to be included in Springer's e-book package.

Typical works might include:

- An extended survey of a field

- A link between new research papers published in journal articles

- A presentation of core concepts that doctoral students must understand in order to make independent contributions

- Lecture notes making a specialist topic accessible for non-specialist readers.

SpringerBriefs in Mathematical Physics showcase, in a compact format, topics of current relevance in the field of mathematical physics. Published titles will encompass all areas of theoretical and mathematical physics. This series is intended for mathematicians, physicists, and other scientists, as well as doctoral students in related areas.

Editorial Board

- Nathanaël Berestycki (University of Cambridge, UK)

- Mihalis Dafermos (University of Cambridge, UK / Princeton University, US)

- Tohru Eguchi (Rikkyo University, Japan)

- Atsuo Kuniba (University of Tokyo, Japan)

- Matilde Marcolli (CALTECH, US)

- Bruno Nachtergaele (UC Davis, US)

SpringerBriefs in a nutshell

Springer

Briefs specifications vary depending on the title. In general, each Brief will have:

- 50–125 published pages, including all tables, figures, and references

- Softcover binding

- Copyright to remain in author's name

- Versions in print, eBook, and MyCopy

More information about this series at http://www.springer.com/series/11953

Marco Stevens

The Kadison-Singer Property

 Springer

Marco Stevens
Section of Analysis, Department
  of Mathematics
KU Leuven
Leuven
Belgium

ISSN 2197-1757   ISSN 2197-1765   (electronic)
SpringerBriefs in Mathematical Physics
ISBN 978-3-319-47701-5   ISBN 978-3-319-47702-2   (eBook)
DOI 10.1007/978-3-319-47702-2

Library of Congress Control Number: 2016954559

Printed on acid-free paper

This Springer imprint is published by Springer Nature
The registered company is Springer International Publishing AG
The registered company address is: Gewerbestrasse 11, 6330 Cham, Switzerland

Foreword

What soon became the *Kadison–Singer conjecture* was formulated by Kadison and Singer in 1959 and was proved (against the negative advice on its validity by the originators!) by Marcus, Spielman, and Srivastava in 2014, after important earlier contributions by Anderson [1], Weaver [2], and others. Despite its seemingly technical setting within operator (algebra) theory, the conjecture and its resolution have generated considerable interest from the mathematical community, as exemplified by, e.g., specialized conferences, a Seminar Bourbaki by Valette, a widely read blog by Tao, coverage by the *Quanta* magazine, and even by the press. This interest may be explained by the unexpectedly large scope of the conjecture (see [3]) as well as by the closely related depth of its proof, which used techniques from diverse fields of mathematics (it may also have helped that Singer shared the 2004 Abel Prize with Atiyah, though for unrelated work).

Despite this interest, a relatively elementary account of the conjecture and its proof was lacking so far. This monograph, which is a revised version of the author's M.Sc Thesis at Radboud University Nijmegen, fills this gap. In fact, it does far more than that; for example, it includes a clean proof that in the so-called continuous case, the conjecture (which indeed was never posited for that case) would be false, which is perhaps as surprising as its truth in the 'discrete case' (see below for this terminology). This was already established by Kadison and Singer themselves, but in a very contrived way. Furthermore, this book contains a detailed proof of the classification of maximal abelian subalgebras of the algebra $B(H)$ of all bounded operators on a separable Hilbert space H that are closed under hermitian conjugation (i.e., MASA's), which lies at the basis of the Kadison–Singer conjecture. There are many other results like those, which make this treatise as complete and self-contained as can be expected given its modest length.

All that remains to be added is a brief account of the historical context of the Kadison–Singer conjecture, which, as the originators acknowledge, was at least in part inspired by quantum mechanics. At the time, the Hilbert space approach to quantum mechanics proposed by von Neumann in 1932 was about 25 years of age. In the meantime, von Neumann, Gelfand, and Naimark had created the new mathematical discipline of operator algebras, to which Kadison (who had been a student of another Hilbert space pioneer, Stone) and Singer's Ph.D. advisor Segal had made important contributions. Moreover, the formalism of quantum mechanics in Hilbert space per se continued to be developed by mathematicians, as exemplified by the famous papers by Gleason [4] and Mackey [5].

Kadison and Singer [6] combined these trends, in analyzing a potential ambiguity in the Hilbert space formalism in terms of operator algebras. To begin with, assume that H is a finite-dimensional Hilbert space, and consider some set $\underline{a} = (a_1, \ldots, a_n)$ of *commuting* self-adjoint operators on H that is *maximal* in the sense that the (commutative) algebra A generated by the operators a_i cannot be extended to some larger *commutative* subalgebra of $B(H)$. Note that A is closed under hermitian conjugation $a \mapsto a^*$; as such, it is called a*-*algebra*. Then, H has an orthonormal basis of joint eigenvectors v_λ of $\underline{a}$, labeled by the joint eigenvalues $\underline{\lambda} = (\lambda_1, \ldots, \lambda_n)$, i.e., $a_i v_\lambda = \lambda_i v_\lambda$. Physicists call unit vectors in Hilbert space 'states,' but in the operator algebra literature, a *state* on some operator algebra $A \subseteq B(H)$ (which for simplicity we assume to contain the unit operator 1_H on H) is defined as a linear map $\omega : A \to \mathbb{C}$ such that: (i) $\omega(a^*a) \geq 0$ for each $a \in A$, and (ii) $\omega(1_H) = 1$. Clearly, each unit vector $|\underline{\lambda}\rangle$ defines a state ω_λ on $B(H)$ by means of

$$\omega_{\underline{\lambda}}(a) = \langle v_{\underline{\lambda}}, a v_{\underline{\lambda}} \rangle,$$

where $\langle , \rangle$ is the inner product in H (note that physicists would write this as something like $\langle a \rangle_{\underline{\lambda}} = \langle \underline{\lambda} | a | \underline{\lambda} \rangle$). This state is *pure*, in being an extreme element of the (compact) convex set of all states on $B(H)$ (i.e., a pure state has no nontrivial decomposition as a convex sum of other states). In fact, as long as $\dim(H) < \infty$, any pure state ω on $B(H)$ takes the form $\omega(a) = \langle \psi, a\psi \rangle$ ($a \in B(H)$), where $\psi \in H$ is some unit vector. By restriction, ω_λ also defines a state on A (which need not be pure). Does its restriction to A conversely determine the original state on $B(H)$?

This question is mathematically non-trivial even for finite-dimensional H (though easy to answer in that case) and is physically interesting for two related reasons. First, the labeling $\underline{\lambda}$ only refers to A, which would make the (Dirac) notation $|\underline{\lambda}\rangle$ (which is meant to define a state on $B(H)$) ambiguous in case the answer to the above question is no. Second, in Bohr's 'Copenhagen Interpretation' of quantum mechanics, both the measurement apparatus and the outcome of any measurement must be recorded in the language of classical physics, which roughly speaking means that the apparatus is mathematically represented by some commutative subalgebra $A \subseteq B(H)$, whereas the outcome (assumed sharp, i.e., dispersion-free) defines a pure state on A. The question, then, is whether such a measurement outcome also fixes the state of the quantum system as a whole.

In the finite-dimensional case, it is easy to show that any maximal commutative subalgebra A of $B(H) \cong M_n(\mathbb{C})$ is (unitarily) conjugate to the algebra of diagonal matrices $D_n(\mathbb{C})$, from which in turn it is straightforward to show that any *pure* state on A indeed has a unique extension to a *pure* state on $M_n(\mathbb{C})$. So everything is fine in that case.

The infinite-dimensionality of H leads to a number of new phenomena:

- There exist pure states ω on $B(H)$ that are *not* represented by any unit vector ψ; such pure states are called *singular* (as opposed to *normal*).
- There exist maximal abelian *-algebras in $B(H)$ that are *not* (unitarily) conjugate.

To proceed, Kadison and Singer assumed that H is *separable*, in having a countable orthonormal basis. In that case, von Neumann himself had already classified the possible maximal abelian *-algebra $A \subseteq B(H)$ up to unitary equivalence, with the result (proved in detail in this book) that A must be equivalent to exactly one of the following:

1. $A_c = L^\infty(0,1) \subseteq B(L^2(0,1))$, called the *continuous case*;
2. $A_d = \ell^\infty(\mathbb{N}) \subseteq B(\ell^2(\mathbb{N}))$, called the *discrete case*; and
3. $A_\kappa = L^\infty(0,1) \oplus \ell^\infty(\kappa) \subseteq B(L^2(0,1) \oplus \ell^2(\kappa))$, called the *mixed case*,

where either $\kappa = \{1,\ldots,n\}$, in which case one has $\ell^2(\kappa) \cong \mathbb{C}^n$ with $\ell^\infty(\kappa) \cong D_n(\mathbb{C})$, or $\kappa = \mathbb{N}$ (the inclusions are given by realizing each commutative algebra by multiplication operators).

In all cases, *normal* pure states on A uniquely extend to (necessarily normal) pure states on $B(H)$. As already mentioned, Kadison and Singer already showed that A_c has singular pure states whose extension to $B(H)$ is far from unique (in fact, every singular pure state on A_c has this property), which also settles the mixed case (i.e., in the negative).

This leaves the discrete case, about which the *Kadison–Singer conjecture* claims that every pure state on $\ell^\infty(\mathbb{N})$ has a unique extension to a pure state on $\ell^2(\mathbb{N})$. So this conjecture is now a theorem, and the best way to find out about it is to continue reading.

Nijmegen   Klaas Landsman
August 2016

References

1. Anderson, J.: Extensions, restrictions and representations of states on C*-algebras. Trans. Am. Math. Soc. **249**(2), 303–329 (1979)
2. Weaver, N.: The Kadison-Singer problem in discrepancy theory. Disc. Math. (278), 227–239 (2004)
3. Casazza, P., Tremain, J.: The Kadison-Singer problem in mathematics and engineering. Proc. Natl. Acad. Sci. **103**(7), 2032–2039 (2006)
4. Gleason, A.: Measures on the closed subspaces of a Hilbert space. J. Math. Mech. (6), 885–893 (1957)
5. Mackey, G.: Quantum mechanics and Hilbert space. Am. Math. Monthly (64), 45–57 (1957)
6. Kadison, R.V., Singer, I.: Extensions of pure states. Am. J. Math. **81**(2), 383–400 (1959)

Contents

Chapter 1
Introduction

In 1959, Richard Kadison and Isadore Singer published the article *'Extensions of pure states'* [1] where they formulated the following question: given a Hilbert space H and a maximal abelian subalgebra A of the operator algebra $B(H)$, does every pure state on A extend to a unique pure state on $B(H)$? In their article, they showed that this question was only open for one algebra: $\ell^\infty(\mathbb{N})$, considered as a subalgebra of $B(\ell^2(\mathbb{N}))$, realized via the multiplication operator. They were not able to answer the question for this algebra, but believed the answer was negative.

This question became known as the *Kadison-Singer conjecture*. It took 54 years before Adam Marcus, Daniel Spielman and Nikhil Srivastava proved [2] that in fact the question had a positive answer for the algebra $\ell^\infty(\mathbb{N})$. For this, they used another conjecture that was formulated in 2004 by Weaver [3], which was already known to imply the Kadison-Singer conjecture. In order to prove Weaver's conjecture, Marcus, Spielman and Srivastava proved two major results involving random variables with matrix values.

In this text, we embed the Kadison-Singer conjecture in the classification of abelian subalgebras with the *Kadison-Singer property*. In Chap. 2, we introduce the concept of pure state extensions by means of a concrete example, namely within the context of a matrix algebra with the algebra of diagonal matrices as a subalgebra. For this finite dimensional case, we can describe states and pure states explicitly and show that any pure state on the diagonal matrices can be uniquely extended to a pure state on the whole matrix algebra.

In Chap. 3, we generalize the concept of states on matrix algebras to states on C*-algebras. Compared to Chap. 2, the role of the matrix algebra is played by the operator algebra $B(H)$, where H is some Hilbert space, and the subalgebra of diagonal matrices is replaced by some abelian C*-subalgebra $A \subseteq B(H)$. Then again, we pose the question: does every pure state on the subalgebra extend uniquely to a pure state on the whole operator algebra? If it does, we say the subalgebra has the *Kadison-Singer property*.

In the rest of the text, we try to classify all abelian subalgebras with the Kadison-Singer property. In Chap. 4, we show that an abelian subalgebra with the Kadison-Singer property is necessarily *maximal abelian*. At this point, we can

© The Author(s) 2016
M. Stevens, *The Kadison-Singer Property*,
SpringerBriefs in Mathematical Physics 14, DOI 10.1007/978-3-319-47702-2_1

appreciate the question of Kadison and Singer in its natural context. In the same chapter, we also give three main examples of maximal abelian subalgebras: the discrete, continuous and mixed subalgebra. Here, the discrete subalgebra can be seen as the proper generalization of the algebra of diagonal matrices.

These three examples are all subalgebras of an operator algebra $B(H)$, where H is separable. In Chap. 5 we show that we only have to consider these examples when considering separable Hilbert spaces, since for these Hilbert spaces, every maximal abelian subalgebra $A \subseteq B(H)$ is unitarily equivalent to one of these three examples. We prove this using the arguments used by Kadison and Ringrose [4], which are based on the concept of minimal projections.

In Chaps. 6, 7 and 8, we complete the classification of abelian subalgebras with the Kadison-Singer property in the separable case. First of all, in Chap. 6 we introduce the concept of ultrafilters and show that we can construct the Stone-Čech compactification of Tychonoff spaces using ultrafilters on zero-sets. We use this in Chap. 7, to show that the continuous subalgebra does not have the Kadison-Singer property, based on the work of Anderson [5]. As a consequence of this, the mixed subalgebra does not have the Kadison-Singer property either.

By then, it is clear that Kadison-Singer conjecture is the only question left in order to complete the classification. In Chap. 8, we first discuss the results of Marcus, Spielman and Srivastava. After that, we prove Weaver's theorem and use this to prove the Kadison-Singer conjecture. For this, we use the adaptation of these results as formulated by Tao [6].

In the appendices, we give some extra material. Appendices A and B provide background knowledge, where appendix A contains a broad range of preliminaries and appendix B is focussed on functional analysis and operator algebras. Appendix C contains some further results that rely on concepts introduced in the main text, but that are at the same time also needed to prove some results there. They are not included in the main text themselves, since they would only distract from the main results there. Finally, in appendix D, we have included some notes and remarks on the main text. Especially, we give a survey of the use of existing literature and we discuss in what way we have improved upon these sources.

References

1. Kadison, R.V., Singer, I.: Extensions of pure states. Am. J. Math. **81**(2), 383–400 (1959)
2. Marcus, A., Spielman, D., Srivastava, N.: Interlacing families II: mixed characteristic polynomials and the Kadison-Singer problem. Ann. Math. **182**, 327–350 (2015)
3. Weaver, N.: The Kadison-Singer problem in discrepancy theory. Discrete Math. **278**, 227–239 (2004)
4. Kadison, R.V., Ringrose, J.R.: Fundamentals of the theory of operator algebras. Pure and Applied Mathematics: Advanced Theory, vol. 2. Academic Press (1986)
5. Anderson, J.: Extensions, restrictions and representations of states on C*-algebras. Trans. Am. Math. Soc. **249**(2), 303–329 (1979)
6. Tao, T.: Real stable polynomials and the Kadison-Singer problem (2013). https://terrytao. wordpress.com/2013/11/04/real-stable-polynomials-and-the-kadison-singer-problem/

Chapter 2
Pure State Extensions in Linear Algebra

In this chapter we introduce the concept of a pure state extension by means of a concrete example: we consider the matrix algebra

$$M := M_n(\mathbb{C}),$$

for some fixed $n \in \mathbb{N}$. We often denote an element $a \in M$ by

$$a = \sum_{i,j} a_{ij} |e_i\rangle \langle e_j|,$$

where $\{e_i\}$ is the standard basis of $\mathbb{C}^n$ and we use the shorthand notation $|x\rangle \langle y|$ for the operator which satisfies $|x\rangle \langle y| (z) = \langle y, z\rangle x$. This means that a_{ij} is the element in the i-th row and j-th column of the matrix a. Furthermore, we consider the diagonal matrices

$$D := \{a \in M \,|\, a_{ij} = 0 \text{ if } i \neq j\},$$

which form a unital subalgebra of M.

The algebra M also has a $*$-operation that is an involution, defined by:

$$a^* = \sum_{i,j} \overline{a_{ji}} |e_i\rangle \langle e_j|.$$

We call a^* the **adjoint** of a. Note that D is also closed under this operation.

© The Author(s) 2016
M. Stevens, *The Kadison-Singer Property*,
SpringerBriefs in Mathematical Physics 14, DOI 10.1007/978-3-319-47702-2_2

2.1 Density Operators and Pure States

M is not merely an algebraic object; it also has its defining action on $\mathbb{C}^n$, which is a vector space with a natural inner product $\langle x, y \rangle = \sum_i \overline{x_i} y_i$ (i.e. we take the standard inner product that is linear in the second argument). Using this, we can define a special class of matrices.

Definition 2.1 $a \in M$ is called **positive** if for each $x \in \mathbb{C}^n$ we have $\langle x, ax \rangle \geq 0$. We write this condition as $a \geq 0$.

Now we can define our main object of study: states.

Definition 2.2 A **state** on M is a linear map $f : M \to \mathbb{C}$ that is positive, meaning that $f(a) \geq 0$ for all $a \geq 0$, and unital, i.e. $f(1) = 1$. The set of all states on M is denoted by $S(M)$, which we call the **state space** of M.

In turns out that all states on M are of a specific form. To make this more precise, we need two more definitions.

Definition 2.3 The **trace** of a matrix $a \in M$ is defined as $\text{Tr}(a) = \sum_i a_{ii}$.

Lemma 2.4 *1. $\text{Tr}(ab) = \text{Tr}(ba)$ for all $a, b \in M$*
2. For any basis $\{v_i\}$ of $\mathbb{C}^n$, we have $\text{Tr}(a) = \sum_i \langle v_i, av_i \rangle$

Proof 1. $\text{Tr}(ab) = \sum_i (ab)_{ii} = \sum_i \sum_k a_{ik} b_{ki} = \sum_k \sum_i b_{ki} a_{ik} = \sum_k (ba)_{kk} = \text{Tr}(ba)$.

2. Note that by definition, $\text{Tr}(a) = \sum_i \langle e_i, ae_i \rangle$. For another basis $\{v_i\}$ there is a unitary $u \in M$, i.e. $uu^* = u^*u = 1$, such that $ue_i = v_i$ for all i. Then:

$$\sum_i \langle v_i, av_i \rangle = \sum_i \langle ue_i, aue_i \rangle = \sum_i \langle e_i, u^*aue_i \rangle = \text{Tr}(u^*au) = \text{Tr}(auu^*) = \text{Tr}(a),$$

where we used part 1 of this lemma. $\square$

There is a connection between states on M and so-called *density operators*.

Definition 2.5 A **density operator** $\rho \in M$ is a positive operator that satisfies $\text{Tr}(\rho) = 1$. We write $\mathscr{D}(M)$ for the set of all density operators in M.

Theorem 2.6 *There is a bijective correspondence between states f on M and density operators $\rho \in M$, given by $f(a) = \text{Tr}(\rho a)$ for all $a \in M$.*

Proof We prove that $S(M) \cong \mathscr{D}(M)$ as sets. We construct $\Phi : S(M) \to \mathscr{D}(M)$ via

$$\Phi(f) = \sum_{i,j} \rho_{ij} |e_i\rangle \langle e_j|,$$

where $\rho_{ij} = f(|e_j\rangle \langle e_i|)$.

To see that Φ is well defined, note that

$$\text{Tr}(\Phi(f)) = \sum_i f(|e_i\rangle\langle e_i|) = f(\sum_i |e_i\rangle\langle e_i|) = f(1) = 1$$

and for $x \in \mathbb{C}^n$, say $x = \sum_i c_i e_i$,

$$\langle x, \Phi(f)x\rangle = \sum_{i,j} \overline{c_i}c_j\langle e_i, \Phi(f)e_j\rangle = \sum_{i,j} \overline{c_i}c_j f(|e_j\rangle\langle e_i|) = f(|x\rangle\langle x|) \geq 0,$$

which means that $\Phi(f)$ is indeed a density operator.

Next, define $\Psi : \mathscr{D}(M) \to S(M)$ by

$$\Psi(\rho)(a) = \text{Tr}(\rho a)$$

for all $a \in M$.

To see that Ψ is well defined, first note that $\Psi(\rho)(1) = \text{Tr}(\rho) = 1$. Next, let $\rho \in \mathscr{D}(M)$ and $a \in M$ positive. Then ρ has a spectral decomposition

$$\rho = \sum_i p_i |v_i\rangle\langle v_i|,$$

for some orthonormal basis (v_i), where all $p_i \geq 0$. Since a is positive,

$$a = \sum_{i,j} \lambda_{ij} |v_i\rangle\langle v_j|,$$

with all $\lambda_{ii} \geq 0$. Then $\rho a = \sum_{i,j} p_i \lambda_{ij} |v_i\rangle\langle v_j|$, so

$$\Psi(\rho)(a) = \text{Tr}(\rho a) = \sum_i p_i \lambda_{ii} \geq 0,$$

so $\Psi(\rho)$ is positive, and hence a state. Now, note that

$$\Psi(\Phi(f))(a) = \text{Tr}(\Phi(f)a) = \text{Tr}((\sum_{i,j} \rho_{ij}|e_i\rangle\langle e_j|)(\sum_{l,k} a_{lk}|e_l\rangle\langle e_k|))$$

$$= \sum_{i,j} \rho_{ij}a_{ji} = \sum_{i,j} a_{ji}f(|e_j\rangle\langle e_i|) = f(\sum_{i,j} a_{ji}|e_j\rangle\langle e_i|)$$

$$= f(a),$$

meaning that $\Psi \circ \Phi = \text{Id}$.

Next,

$$\Phi(\Psi(\rho))_{ij} = \Psi(\rho)(|e_j\rangle\langle e_i|) = \text{Tr}(\rho|e_j\rangle\langle e_i|) = \langle e_i, \rho e_j\rangle = \rho_{ij},$$

meaning that $\Phi \circ \Psi = \mathrm{Id}$. Hence, $\mathscr{D}(M) \cong S(M)$ as sets, and writing $\Psi(\rho) = f$ the given formula $f(a) = \mathrm{Tr}(\rho a)$ holds. $\square$

Note that $S(M)$ and $\mathscr{D}(M)$ have more structure than that of a set, since they are also convex. A function $f : A \to B$ between two convex sets is called **affine** if it preserves the convex structure, i.e. if $f(tx + (1-t)y) = tf(x) + (1-t)f(y)$ for all $t \in [0, 1]$ and $x, y \in A$. Note that the bijection in Theorem 2.6 is an affine function.

For a convex set C, a point $c \in C$ is called **extreme** if for any $c_1, c_2 \in C$ and $t \in (0, 1)$ such that $c = tc_1 + (1-t)c_2$ we have $c_1 = c_2 = c$. The set of extreme points of a convex set C is called the **extreme boundary** of C, often denoted as $\partial_e C$.

Since $S(M)$ is a convex set, we can consider its boundary, which plays a crucial role in our discussion. For the elements in this boundary, i.e. the extreme points of $S(M)$, we have a special name.

Definition 2.7 A state $f \in S(M)$ is called a **pure state** if it is an extreme point of $S(M)$.

To determine the pure states on M, we use the following lemma.

Lemma 2.8 *Suppose that C and D are convex sets and that there is an affine isomorphism between them. Then $\partial_e C$ is isomorphic to $\partial_e D$.*

Proof Suppose that the map $\phi : C \to D$ is an affine isomorphism. First of all, we claim that $\phi(\partial_e C) \subseteq \partial_e D$.

To see this, first note that ϕ^{-1} is an affine isomorphism as well. Now suppose $x \in \partial_e C$ and $t \in [0, 1]$, $a, b \in D$ such that $\phi(x) = ta + (1-t)b$. Then

$$x = \phi^{-1}(ta + (1-t)b) = t\phi^{-1}(a) + (1-t)\phi^{-1}(b).$$

Then, since $x \in \partial_e C$, $x = \phi^{-1}(a) = \phi^{-1}(b)$, but then also $\phi(x) = a = b$, so we have that $\phi(x) \in \partial_e D$.

Hence $\phi(\partial_e C) \subseteq \partial_e D$, so by the same token $\phi^{-1}(\partial_e D) \subseteq \partial_e C$, whence ϕ maps $\partial_e C$ bijectively to $\partial_e D$. Therefore $\partial_e C$ and $\partial_e D$ are isomorphic. $\square$

We can now give an explicit description of the pure states on M.

Corollary 2.9 *There is a bijective correspondence between pure states f on M and one-dimensional projections $|\psi\rangle \langle\psi|$, such that $f(a) = \langle\psi, a\psi\rangle$ for all $a \in M$.*

Proof By Theorem 2.6 we know that $S(M)$ corresponds bijectively to $\mathscr{D}(M)$ via the formula $f(a) = \mathrm{Tr}(\rho a)$. Since this formula is affine and the pure states on M are exactly $\partial_e S(M)$, we only need to determine $\partial_e \mathscr{D}(M)$, by Lemma 2.8.

Suppose that $\rho \in \partial_e \mathscr{D}(M)$ and let $\rho = \sum_i p_i |v_i\rangle \langle v_i|$ be its spectral decomposition. Then since ρ is positive and has unit trace, we know that the $\{v_i\}$ are orthonormal, all $p_i \geq 0$ and $\sum_i p_i = 1$. Clearly, all $p_i \in [0, 1]$.

Now suppose that there is a $j \in \{1, \ldots, n\}$ such that $p_j \in (0, 1)$. Then there must be a $k \neq j$ such that $p_k \in (0, 1)$ as well. Then there is a $\varepsilon > 0$ such that we have $[p_j - \varepsilon, p_j + \varepsilon] \subseteq [0, 1]$ and $[p_k - \varepsilon, p_k + \varepsilon] \subseteq [0, 1]$. Now define

$$r_i = \begin{cases} p_i - \varepsilon : i = j \\ p_i + \varepsilon : i = k \\ p_i \quad\;\; : i \notin \{j, k\} \end{cases}$$

and

$$q_i = \begin{cases} p_i + \varepsilon : i = j \\ p_i - \varepsilon : i = k \\ p_i \quad\;\; : i \notin \{j, k\}. \end{cases}$$

By construction, $\rho_1 := \sum_i r_i |v_i\rangle \langle v_i|$ and $\rho_2 := \sum_i q_i |v_i\rangle \langle v_i|$ are density operators too, and $\rho = \frac{1}{2}\rho_1 + \frac{1}{2}\rho_2$. However, $\rho_1 \neq \rho \neq \rho_2$, so ρ is not an extreme point of $\mathscr{D}(M)$. Contradiction, since $\rho \in \partial_e \mathscr{D}(M)$ by assumption. Therefore, all $p_i \in \{0, 1\}$. Combined with $\sum_i p_i = 1$, this gives a unique j such that $p_j = 1$ and $p_k = 0$ for all $k \neq j$. But then, $\rho = |v_j\rangle\langle v_j|$, so we see that every extreme point of $\mathscr{D}(M)$ is indeed a one-dimensional projection.

It is clear that every one-dimensional projection is positive and has unit trace, so every one-dimensional projection is clearly a density operator. Now take a one-dimensional projection $\rho = |\psi\rangle \langle \psi|$, i.e. a unit vector ψ. Suppose that there are $\rho_1, \rho_2 \in \mathscr{D}(M)$ and a $t \in (0, 1)$ such that $\rho = t\rho_1 + (1 - t)\rho_2$.

Clearly, we have $\langle \psi, \rho\psi \rangle = 1$. Using the spectral decomposition $\sum_i p_i |v_i\rangle \langle v_i|$ of ρ_1, where the $\{v_i\}$ are orthonormal, all $p_i \geq 0$ and $\sum_i p_i = 1$, we see that:

$$\langle \psi, \rho_1 \psi \rangle = \sum_i p_i |\langle \psi, v_i \rangle|^2 \leq \sum_i p_i = 1,$$

by the Cauchy-Schwarz inequality.

By the same token, $\langle \psi, \rho_2 \psi \rangle \leq 1$. Therefore,

$$1 = \langle \psi, \rho\psi \rangle = t\langle \psi, \rho_1\psi \rangle + (1 - t)\langle \psi, \rho_2\psi \rangle \leq t + (1 - t) = 1.$$

Therefore, we must have $\langle \psi, \rho_1\psi \rangle = 1$, so for all j such that $p_j \neq 0$, we have $|\langle \psi, v_j \rangle|^2 = 1$. Since ψ is a unit vector and $\{v_i\}$ is an orthonormal set, this means that there is a unique j such that $p_j \neq 0$ and $\psi = zv_i$ with $z \in \mathbb{C}$ such that $|z| = 1$.

But then necessarily $p_j = 1$ and $\rho_1 = |v_j\rangle\langle v_j| = |\psi\rangle \langle \psi| = \rho$. Likewise, $\rho_2 = \rho$, so indeed, ρ is an extreme point.

So $\partial_e \mathscr{D}(M)$ consists exactly of the one-dimensional projections. Now, under the correspondence of Theorem 2.6,

$$f(a) = \mathrm{Tr}(|\psi\rangle \langle \psi| a) = \langle \psi, |\psi\rangle \langle \psi| a\psi \rangle = \langle \psi, a\psi \rangle,$$

where we used an orthonormal basis with ψ as one of the basis vectors for evaluating the trace. $\qquad\square$

In the same fashion we can also define (pure) states on D and derive their specific forms. Note that for $a \in D$ the notion of positivity when considering it as an element of M, i.e. $\langle x, ax \rangle \geq 0$ for all $x \in \mathbb{C}^n$, is equivalent to saying that all $a_{ii} \geq 0$.

Definition 2.10 A **state** on D is a linear function $f : D \to \mathbb{C}$ that is positive and unital, meaning that $f(a) \geq 0$ for all $a \geq 0$ and $f(1) = 1$. The set of all states on D is denoted by $S(D)$ and is called the **state space** of D.

In our discussion about the the specific form of states on D, we need (to repeat) the notion of a probability distribution on finite sets.

Definition 2.11 Let X be a finite set. Then a **probability distribution** on X is a function $p : X \to [0, \infty)$ such that $\sum_x p(x) = 1$. The set of all probability distributions on X is denoted by $Pr(X)$.

Note that a probability distribution p on a finite set X is equivalently defined as a map $p : X \to [0, 1]$ such that $\sum_x p(x) = 1$.

Theorem 2.12 *There is a bijective correspondence between states f on D and probability distributions p on $\{1, \ldots, n\}$ such that $f(a) = \sum_i p(i)a_{ii}$ for all $a \in D$.*

Proof We want to show that $S(D) \cong Pr(\{1, \ldots, n\})$ as sets.

Define $\Phi : S(D) \to Pr(\{1, \ldots, n\})$ by

$$\Phi(f)(k) = f(|e_k\rangle \langle e_k|)$$

for all k. Then since f is a state, each $\Phi(f)(k)$ is positive. Furthermore,

$$\sum_i \Phi(f)(i) = \sum_i f(|e_i\rangle \langle e_i|) = f(\sum_i |e_i\rangle \langle e_i|) = f(1) = 1,$$

so $\Phi(f)$ is indeed a probability distribution. Next, define the function $\Psi : Pr(\{1, \ldots, n\}) \to S(D)$ by

$$\Psi(p)(a) = \sum_i p(i)a_{ii}.$$

Since all $p(i)$ are positive, it is clear that $\Psi(p)$ is positive too. Furthermore,

$$\Psi(p)(1) = \sum_i p(i) = 1,$$

so $\Psi(p)$ is indeed a state. Now note that

$$\Psi(\Phi(f))(a) = \sum_i \Phi(f)(i)a_{ii} = \sum_i a_{ii} f(|e_i\rangle \langle e_i|) = f(\sum_i a_{ii} |e_i\rangle \langle e_i|) = f(a),$$

showing that $\Psi \circ \Phi = \text{Id}$.

Furthermore,

$$\Phi(\Psi(p))(k) = \Psi(p)(|e_k\rangle \langle e_k|) = \sum_i p(i)(|e_k\rangle \langle e_k|)_{ii} = p(k),$$

whence $\Phi \circ \Psi = \mathrm{Id}$.

So, indeed, $S(D) \cong Pr(\{1, \ldots, n\})$ as sets and writing $p = \Phi(f)$, the given formula $f(a) = \sum_i p(i)a_{ii}$ holds for every $a \in D$. $\qquad\qquad\qquad$ $\square$

Next, we note that just like in the case of M, the state space $S(D)$ is in fact a convex set, just like $Pr(\{1, \ldots, n\})$. Hence we can again determine the boundary of $S(D)$ and call it the **pure state space** of D. Once again, these pure states have a specific form.

Corollary 2.13 *For every pure state f on D there is an $i \in \{1, \ldots, n\}$ such that $f(a) = a_{ii}$ for all $a \in D$.*

Proof By Theorem 2.12 we know that the states on D correspond to $Pr(\{1, \ldots, n\})$, and by Lemma 2.8 we then know that we only have to determine the boundary of $Pr(\{1, \ldots, n\})$. If we show that these are exactly those probability distributions that have a unique j such that $p(j) = 1$ and $p(k) = 0$ for all $k \neq j$, we are done.

So, suppose that $p \in \partial_e Pr(\{1, \ldots, n\})$. By definition of a probability distribution, we have $p(j) \in [0, 1]$ for all j. Suppose that $p(j) \in (0, 1)$ for some j. Then there must be a $k \neq j$ such that $p(k) \in (0, 1)$ as well. Then there is a $\varepsilon > 0$ such that

$$[p(j) - \varepsilon, p(j) + \varepsilon] \subseteq [0, 1]$$

and

$$[p(k) - \varepsilon, p(k) + \varepsilon] \subseteq [0, 1].$$

By the same reasoning as in the proof of Corollary 2.9, p is not an extreme point. Contradiction. Hence there is no j such that $p(j) \in (0, 1)$, so all $p(j) \in \{0, 1\}$. Therefore, there is a unique j such that $p(j) = 1$ and $p(k) = 0$ for all $k \neq j$.

Now suppose p is a probability distribution such that there is a unique j such that $p(j) = 1$ and $p(k) = 0$ for all $k \neq j$. Then suppose that we have a $t \in (0, 1)$ and $r, q \in Pr(\{1, \ldots, n\})$ such that $p = tr + (1 - t)q$. Suppose that $r(j) \neq 1$. Then $r(j) < 1$, since all $r(k) \geq 0$ and $\sum_k r(k) = 1$. Then $q(j) > 1$, which is a contradiction. Hence $r(j) = 1$. Likewise, $q(j) = 1$. Then, since $r, q \in Pr(\{1, \ldots, n\})$, $r(k) = 0 = q(k)$ for all $k \neq j$. Therefore $p = q = r$ and p is an extreme point. $\quad\square$

2.2 Extensions of Pure States

We have now established the ingredients to get to the main point of this chapter. By definition of the state spaces, it is clear that when restricting a state on M one obtains a state on D. The question we can now ask ourselves is whether this restriction determines the original state completely, i.e. whether we can *uniquely extend* a state on D to a state on M. It turns out that it does when we consider pure states, as formulated in the following theorem.

Theorem 2.14 *For every pure state f on D there is a unique pure state g on M that extends f.*

Proof Let f be a pure state on D. By Corollary 2.13, there is an $i \in \{1, \ldots, n\}$ such that $f(a) = a_{ii}$ for all $a \in D$.

Now simply define the linear function $g : M \to \mathbb{C}$ by

$$g(a) = a_{ii}$$

for all $a \in M$. Then clearly, $g(a) = a_{ii} = \langle e_i, ae_i \rangle \geq 0$ for all $a \geq 0$, so g is positive. Furthermore, g is obviously unital, so g is a state that extends f.

Suppose that g' is another pure state that extends f. Then by Corollary 2.9, there is a $\psi \in \mathbb{C}^n$ such that $g'(a) = \langle \psi, a\psi \rangle$ for all $a \in M$.

Let us write $\psi = \sum_k c_k e_k$. Then, since $|e_k\rangle \langle e_k| \in D$ for all k:

$$|c_k|^2 = g'(|e_k\rangle \langle e_k|) = f(|e_k\rangle \langle e_k|) = \delta_{ik}$$

Therefore, $\psi = c_i e_i$, with $|c_i| = 1$. Then for any $a \in M$,

$$g'(a) = \langle \psi, a\psi \rangle = \overline{c_i} c_i \langle e_i, ae_i \rangle = |c_i|^2 a_{ii} = a_{ii} = g(a),$$

so $g' = g$, and g is the unique pure state extension of f. $\square$

Chapter 3
State Spaces and the Kadison-Singer Property

In Chap. 2 we discussed the extension of pure states from the algebra of diagonal matrices D to the algebra of matrices M. In this chapter, we formulate the question whether this is possible in a much broader setting. Instead of M we consider $B(H)$ for some Hilbert space H, and instead of D we consider abelian C^*-subalgebras A of $B(H)$. Having again defined (pure) states, we will likewise ask the question whether a unique extension property holds. This property is the so-called Kadison-Singer property.

3.1 States on C*-Algebras

Using the notion of positivity as introduced in Definition B.18, we can define states.

Definition 3.1 Let A be a unital C^*-algebra. A **state** on A is a linear map $f : A \to \mathbb{C}$ that is positive (i.e. $f(a) \geq 0$ for all $a \geq 0$) and unital (i.e. $f(1) = 1$). The set of all states on A is denoted by $S(A)$ and is called the **state space** of A.

The condition of being positive has a very important consequence for states.

Proposition 3.2 *Suppose A is a unital C^*-algebra and $f \in S(A)$. Then*

$$\sup\{|f(a)| : a \in A, \ \|a\| = 1\}$$

is finite, i.e. $S(A) \subseteq A^*$.

Proof First suppose that $\sup\{|f(a)| : \|a\| = 1, a \geq 0\}$ is infinite. Then there is a sequence $\{a_n\}_{n \in \mathbb{N}}$ such that $|f(a_n)| \geq 2^n$, $a_n \geq 0$ and $\|a_n\| = 1$ for all $n \in \mathbb{N}$. Then $a = \sum_{n=1}^{\infty} 2^{-n} a_n$ exists and is positive too. Then, by linearity, $1 \leq f(2^{-n} a_n)$ for all $n \in \mathbb{N}$. Hence we have

© The Author(s) 2016
M. Stevens, *The Kadison-Singer Property*,
SpringerBriefs in Mathematical Physics 14, DOI 10.1007/978-3-319-47702-2_3

$$N \le \sum_{n=1}^{N} f(2^{-n} a_n) = f(\sum_{n=1}^{N} 2^{-n} a_n) \le f(a),$$

i.e. $N \le f(a)$ for all $N \in \mathbb{N}$. This is a contradiction, so

$$M := \sup\{|f(a)| : \|a\| = 1, a \ge 0\}$$

is finite. Now let $a \in A$ be an arbitrary element such that $\|a\| = 1$. Then a can be written as $a = \sum_{k=0}^{3} i^k a_k$ where all $a_k \ge 0$ and $\|a_k\| \le 1$ by Proposition B.20. Therefore,

$$|f(a)| = \left| f(\sum_{k=0}^{3} i^k a_k) \right| = \left| \sum_{k=0}^{3} i^k f(a_k) \right| \le \sum_{k=0}^{3} \|a_k\| |f(\tfrac{a_k}{\|a_k\|})| \le 4M,$$

i.e. $\sup_{\|a\|=1} |f(a)|$ is finite too.   □

When considering states, the following result is often useful.

Lemma 3.3 *Suppose A is a C^*-algebra and $f \in S(A)$. Then the map*

$$A^2 \to \mathbb{C}, (a, b) \mapsto f(a^* b)$$

is a pre-inner product and hence for every $a, b \in A$ we have

$$|f(a^* b)| \le f(a^* a)^{1/2} f(b^* b)^{1/2}.$$

Proof Since f is positive, this is immediate from Corollary A.2 and the Cauchy-Schwarz inequality for pre-inner products.   □

This has the following corollary:

Corollary 3.4 *Suppose A is a unital C^*-algebra and $f \in S(A)$. Furthermore, let $a \in A$. Then $f(a^*) = \overline{f(a)}$.*

Proof We use Lemma 3.3 to see that $f(a^*) = f(a^* 1) = \overline{f(1^* a)} = \overline{f(a)}$.   □

Since every state is bounded by Proposition 3.2, we can consider its norm. Using this, we can give a different characterization of states.

Proposition 3.5 *Suppose that H is a Hilbert space and A is a unital C^*-subalgebra of $B(H)$. Furthermore, let $f : A \to \mathbb{C}$ be a bounded functional such that $f(1) = 1$. Then f is positive (and hence a state) iff $\|f\| = 1$.*

Proof First suppose that f is positive. Since $\|1\| = 1$, $\|f\| \ge |f(1)| = 1$.
    Now let $a \in A$ such that $\|a\| = 1$. Then, using Lemma 3.3,

$$|f(a)|^2 = |f(1a)|^2 \le f(1^* 1) f(a^* a) \le f(1) \|f\| \|a^* a\| = \|f\|$$

Therefore,

$$\|f\|^2 = \sup_{\|a\|=1} |f(a)|^2 \le \|f\|,$$

whence $\|f\| \le 1$. So $\|f\| = 1$.

For the converse, suppose that $\|f\| = 1$. Let $a \in A$ be self-adjoint and let $n \in \mathbb{Z}$. Since $f(a) \in \mathbb{C}$, we can write $f(a) = \alpha + i\beta$, with $\alpha, \beta \in \mathbb{R}$. Furthermore, denote $c := \|a^2\|$.

Then:

$$\begin{aligned}
|f(a+in1)|^2 &\le \|f\|^2 \|a+in1\|^2 = \|(a+in1)^*(a+in1)\| \\
&= \|(a-in1)(a+in1)\| = \|a^2+n^21\| \\
&\le \|a^2\| + n^2\|1\| = c + n^2
\end{aligned}$$

Moreover,

$$\begin{aligned}
|f(a+in1)|^2 &= |f(a)+inf(1)|^2 = |\alpha + i\beta + in|^2 \\
&= \alpha^2 + (\beta+n)^2 = \alpha^2 + \beta^2 + 2\beta n + n^2.
\end{aligned}$$

Collecting this, we obtain the inequality:

$$\alpha^2 + \beta^2 + 2\beta n + n^2 \le c + n^2.$$

Rewriting this, we obtain:

$$2\beta n \le c - \alpha^2 - \beta^2.$$

If $\beta \ne 0$, then we obtain for every $n \in \mathbb{Z}$:

$$n \le \frac{c - \alpha^2 - \beta^2}{2\beta},$$

which is a contradiction since the right hand side is independent of n. Hence $\beta = 0$, so $f(a) = \alpha$, i.e. $f(a)$ is real.

Now let $a \ge 0$, $a \ne 0$ and write $b = \frac{a}{\|a\|}$. Since a is self-adjoint, b is self-adjoint and $\|b\| = 1$. We claim that $1 - b$ is positive. To see this, let $x \in H$ and compute:

$$\langle x, (1-b)x \rangle = \langle x, x \rangle - \langle x, bx \rangle \ge \|x\|^2 - \|x\|\|bx\| \ge \|x\|^2 - \|b\|\|x\|^2 \ge 0.$$

So, indeed $1 - b$ is positive and hence also self-adjoint. Since $0 \le 1 - b \le 1$ we also have $\|1 - b\| \le 1$. Then:

$$1 - f(b) = f(1) - f(b) = f(1-b) \le |f(1-b)| \le \|f\|\|1-b\| \le 1,$$

whence $f(b) \geq 0$. Then also $f(a) = \|a\| f(b) \geq 0$. Since we obviously also have that $f(0) \geq 0$, f is positive. $\square$

Since all states on a unital C*-algebra A are bounded by Proposition 3.2, $S(A)$ inherits the weak*-topology from A^* (see Sect. B.1). With respect to this topology, $S(A)$ has an important property.

Proposition 3.6 *Let A be a unital C*-algebra. Then $S(A) \subseteq A^*$ is a compact Hausdorff space.*

Proof We first claim that $S(A) \subseteq A^*$ is closed with respect to the weak*-topology. To see this, suppose that $\{f_i\}$ is a net of states converging to a certain $f \in A^*$. By the definition of the weak*-topology, this means that $f(a) = \lim f_i(a)$ for all $a \in A$.

So, certainly, when taking $a = 1$, it follows that $f(1) = \lim f_i(1) = \lim 1 = 1$, since every f_i is a state. Furthermore, if $a \geq 0$, then $f_i(a) \geq 0$ for every i, so then $f(a) = \lim f_i(a) \geq 0$ as well. So, indeed, $f \in S(A)$, i.e. $S(A)$ is closed with respect to the weak*-topology on A^*.

Now, by the Banach-Alaoglu theorem (see Theorem B.1), the closed unit ball A_1^* of A^* is compact with respect to the weak*-topology and by Proposition 3.5 $S(A) \subseteq A_1^*$. Hence $S(A)$ is closed with respect to the relative topology on A_1^*, which is a compact space. Hence $S(A)$ is compact with respect to the relative topology and therefore with respect to the weak*-topology.

Next, to see that $S(A)$ is Hausdorff, suppose $f, g \in S(A)$ such that $f \neq g$. Then there is an $a \in A$ such that $f(a) \neq g(a)$. Therefore, $\delta := |f(a) - g(a)| > 0$. Now consider $U = B(f, a, \frac{\delta}{2}) \cap S(A)$ and $V = B(g, a, \frac{\delta}{2}) \cap S(A)$. Then both $U, V \subseteq S(A)$ are open and $f \in U$, $g \in V$. Furthermore, $h \in U \cap V$ implies

$$|f(a) - g(a)| \leq |f(a) - h(a)| + |h(a) - g(a)| < \frac{\delta}{2} + \frac{\delta}{2} = \delta,$$

which is a contradiction. Hence $U \cap V = \emptyset$. Therefore, $S(A)$ is Hausdorff. $\square$

3.2 Pure States and Characters

Just like in Chap. 2, we note that $S(A)$ is convex for every unital C*-algebra A. Therefore, we can again consider its boundary $\partial_e S(A)$ and call this the **pure state space** of A. It turns out that in the case that A is abelian, the pure states are exactly the characters (see Definition B.23). To prove this, we first need an equivalent definition of pure states in terms of positive functionals.

Lemma 3.7 *Suppose H is a Hilbert space and $A \subseteq B(H)$. Furthermore, suppose $f \in S(A)$. Then $f \in \partial_e S(A)$ if and only if for all $g : A \to \mathbb{C}$ such that $0 \leq g \leq f$ we have $g = tf$ for some $t \in [0, 1]$.*

Proof Suppose $f \in \partial_e S(A)$ and $g : A \to \mathbb{C}$ such that $0 \le g \le f$. Since $1 \ge 0$, then $0 \le g(1) \le f(1) = 1$.

Now, there are a few cases. First of all, suppose $g(1) = 0$. Then let $a \in A$ be positive. Then by Lemma B.22, $0 \le \frac{a}{\|a\|} \le 1$, whence $0 \le a \le \|a\|1$. Therefore,

$$0 \le g(a) \le g(\|a\|1) = \|a\|g(1) = 0.$$

Since every $b \in A$ can be written as $b = \sum_{k=0}^{3} i^k b_k$ for some $b_k \ge 0$, $g(b) = 0$ for every $b \in A$, i.e. $g = 0$.

As a second case, suppose $g(1) = 1$. Then $f - g \ge 0$ and $(f - g)(1) = 0$, so by the same reasoning as in the first case, $f - g = 0$, i.e. $g = f$.

Lastly, there is the case $0 < g(1) < 1$. In this case, define two functionals g_1 and g_2 by $g_1 = \frac{1}{1-g(1)}(f - g)$ and $g_2 = \frac{1}{g(1)}g$. Then clearly, g_1 and g_2 are both positive and $g_1(1) = g_2(1) = 1$, so $g_1, g_2 \in S(A)$. Furthermore,

$$(1 - g(1))g_1 + g(1)g_2 = f - g + g = f$$

and $f \in \partial_e S(A)$, so $g_1 = g_2 = f$. Therefore, $g = g(1)g_2 = g(1)f$.

In all cases, we see that $g = g(1)f$, and $g(1) \in [0, 1]$.

For the converse, suppose that for all $g : A \to \mathbb{C}$ such that $0 \le g \le f$ there is a $t \in [0, 1]$ such that $g = tf$. Then suppose that $h_1, h_2 \in S(A)$ and $s \in (0, 1)$ such that $f = sh_1 + (1 - s)h_2$. Then $f - sh_1 = (1 - s)h_2 \ge 0$, so $0 \le sh_1 \le f$. Hence, there is a $t \in [0, 1]$ such that $sh_1 = tf$. However, $s = sh_1(1) = tf(1) = t$, so $h_1 = f$. Then also $h_2 = f$, so $f \in \partial_e S(A)$. $\qquad\qquad\square$

Now we can come to our main point; the pure states are exactly the characters, which are defined as in Definition B.23. In Chap. 2, we already saw that every pure state on D was of the form $f(a) = a_{ii}$, which is clearly multiplicative on the diagonal matrices, i.e. $\partial_e S(D) \subseteq \Omega(D)$. Therefore, the following theorem can be seen as a generalization.

Theorem 3.8 *Suppose H is a Hilbert space and let $A \subseteq B(H)$ be an abelian unital C^*-algebra. Then $\partial_e S(A) = \Omega(A)$.*

Proof First let $f \in \partial_e S(A)$. Let $a, c \in A$ and first suppose that $0 \le c \le 1$. Now let $b \in A$ such that $b \ge 0$.

Then $c = d^*d$, $1 - c = u^*u$ and $b = v^*v$ for some $c, u, v \in A$. Therefore,

$$bc = v^*vd^*d = d^*v^*vd = (vd)^*vd \ge 0$$

and

$$b - bc = b(1 - c) = v^*vu^*u = u^*v^*vu = (vu)^*vu \ge 0,$$

so $0 \le bc \le b$.

Now define $g : A \to \mathbb{C}$ by $g(z) = f(zc)$ for all $z \in A$. Combining the fact that $f \ge 0$ and the above observation that $bc \ge 0$ for all $b \ge 0$, we see that $g \ge 0$.

Furthermore, for $b \geq 0$, $b \geq bc$ and hence

$$(f - g)(b) = f(b) - f(bc) = f(b - bc) \geq 0,$$

so $g \leq f$. Now using Lemma 3.7, we know that $g = tf$ for some $t \in [0, 1]$. Now

$$f(ac) = g(a) = tf(a) = tf(1)f(a) = g(1)f(a) = f(c)f(a) = f(a)f(c).$$

If we now drop the requirement that $0 \leq c \leq 1$, we observe that we still have $c = \sum_{k=0}^{3} i^k c_k$ for some $c_k \geq 0$, by Proposition B.20.
Then $c = \sum_{k=0}^{3} i^k \|c_k\| \frac{c_k}{\|c_k\|}$ and $0 \leq \frac{c_k}{\|c_k\|} \leq 1$ by Lemma B.22, whence

$$f(ac) = f\left(a \sum_{k=0}^{3} i^k \|c_k\| \frac{c_k}{\|c_k\|}\right) = \sum_{k=0}^{3} i^k \|c_k\| f\left(a \frac{c_k}{\|c_k\|}\right)$$

$$= \sum_{k=0}^{3} i^k \|c_k\| f(a) f\left(\frac{c_k}{\|c_k\|}\right) = f(a) f\left(\sum_{k=0}^{3} i^k \|c_k\| \frac{c_k}{\|c_k\|}\right)$$

$$= f(a)f(c),$$

i.e. $f \in \Omega(A)$, since $f(1) = 1$ and hence $f \neq 0$. Therefore $\partial_e S(A) \subseteq \Omega(A)$.
For the converse, suppose $c \in \Omega(A)$. Then $c(1) = 1$ by Lemma B.24. Furthermore, for $a \in A$, by Lemma B.24,

$$c(a^* a) = c(a^*)c(a) = \overline{c(a)}c(a) = |c(a)|^2 \geq 0,$$

so $c \geq 0$. Since c is also linear, $c \in S(A)$.
Now we claim that in fact $c \in \partial_e S(A)$. To see this, suppose that $t \in (0, 1)$ and $c_1, c_2 \in S(A)$ such that $c = tc_1 + (1 - t)c_2$. Furthermore, suppose that $a = a^* \in A$. Then $c_1(a) \in \mathbb{R}$, since $c_1 \geq 0$ and $c_1(a)^2 = |c_1(1^*a)|^2 \leq c_1(1^*1)c_1(a^*a) = c_1(a^2)$. Likewise, $c_2(a)^2 \leq c_2(a^2)$.
Since c is a character, we can compute:

$$0 = c(a^2) - c(a)^2$$

$$= tc_1(a^2) + (1 - t)c_2(a^2) - (tc_1(a) + (1 - t)c_2(a))^2$$

$$= tc_1(a^2) + (1 - t)c_2(a^2) - t^2 c_1(a)^2 - (1 - t)^2 c_2(a)^2 - 2t(1 - t)c_1(a)c_2(a)$$

$$\geq tc_1(a)^2 + (1 - t)c_2(a)^2 - t^2 c_1(a)^2 - (1 - t)^2 c_2(a)^2 - 2t(1 - t)c_1(a)c_2(a)$$

$$= (t - t^2)c_1(a)^2 + ((1 - t) - (1 - t)^2)c_2(a)^2 - 2t(1 - t)c_1(a)c_2(a)$$

$$= t(1 - t)(c_1(a)^2 + c_2(a)^2 - 2c_1(a)c_2(a))$$

$$= t(1 - t)(c_1(a) - c_2(a))^2 \geq 0,$$

i.e. $c_1(a) = c_2(a)$ for all $a = a^* \in A$. Therefore, for any $b \in A$, $b = a_1 + ia_2$ with $a_1 = a_1^*, a_2 = a_2^* \in A$, whence $c_1(b) = c_2(b)$ by linearity. Therefore $c_1 = c_2 = c$ and $c \in \partial_e S(A)$. $\qquad\square$

The above theorem is remarkable, since the algebra $B(H)$ for a Hilbert space H of dimension at least 2 does not even admit any characters. This follows directly from the fact that $B(H)$ is non-commutative in this case.

Furthermore, Theorem 3.8 has the following corollary.

Corollary 3.9 *Suppose A is an abelian unital C^*-algebra. Then $\partial_e S(A)$ is compact Hausdorff with respect to the weak*-topology.*

Proof Since $\partial_e S(A) \subseteq S(A)$ and $S(A)$ is Hausdorff, we know that $\partial_e S(A)$ is Hausdorff too. In fact, we only need to show that $\Omega(A) = \partial_e S(A)$ is closed in $S(A)$, since $S(A)$ is compact by Proposition 3.6. To prove this, we show that $U := S(A) \setminus \Omega(A)$ is open in $S(A)$. For this, suppose $f \in U$. Then there are $a, b \in A$ such that $f(a)f(b) \neq f(ab)$. Since every element of A can be written as a sum of positive elements (see Proposition B.20) we know that we can then assume that a and b are positive.

Now, since A is abelian we then also know that ab is positive. Hence $f(a), f(b)$ and $f(ab)$ are positive numbers. If we now suppose that $f(a)f(b) > f(ab)$, we can define $\delta = f(a)f(b) - f(ab) > 0$. Next, define $\varepsilon_1 = \frac{\delta}{f(a)+f(b)+1}$. Using this, we define $\varepsilon = \min\{\varepsilon_1, f(a), f(b)\} > 0$.

Then, take $g \in B(f, a, \varepsilon) \cap B(f, b, \varepsilon) \cap B(f, ab, \varepsilon) \cap S(A)$. Then we have

$$
\begin{aligned}
g(a)g(b) - g(ab) &> (f(a) - \varepsilon)(f(b) - \varepsilon) - (f(ab) + \varepsilon) \\
&= f(a)f(b) - f(ab) - \varepsilon(f(a) + f(b) + 1) + \varepsilon^2 \\
&> \delta - \varepsilon(f(a) + f(b) + 1) \\
&\geq \delta - \delta = 0,
\end{aligned}
$$

i.e. $g(a)g(b) \neq g(ab)$. Hence $g \in U$. A similar argument works if $f(a)f(b) < f(ab)$. Hence U is open. Therefore, $\partial_e S(A) = \Omega(A) \subseteq S(A)$ is closed and hence a compact Hausdorff space. $\qquad\square$

3.3 Extensions of Pure States

Recall that our goal is to generalize the concept of the extension of pure states from the algebra of diagonal matrices D to the algebra of all matrices, M. We have already generalized $D \subseteq M$ to $A \subseteq B(H)$ for a Hilbert space H and an *abelian* unital C^*-subalgebra A. In this case it is important to note that the pure states on A are in fact characters. These cannot be extended to characters on all of $B(H)$, since the latter do not exist. However, they might be extended to states on all of $B(H)$. The question whether this is possible is the one we are interested in.

Definition 3.10 Let H be a Hilbert space and A an abelian unital C^*-subalgebra of $B(H)$. Furthermore, let $f \in S(A)$. We define the **set of extensions of** f to be:

$$\mathrm{Ext}(f) = \{g \in S(B) : g|_A = f\}.$$

In Chap. 2 we showed that for the case $H = \mathbb{C}^n$ and $A = D$, for each $f \in \partial_e S(D)$ the set $\mathrm{Ext}(f) \cap \partial_e S(M)$ consists of exactly one element, i.e. every pure state on D extends to a unique pure state on M. This motivates the following definition.

Definition 3.11 Let H be a Hilbert space and A an abelian unital C^*-subalgebra of $B(H)$. We say that A has the **first Kadison-Singer property** if for every $f \in \partial_e S(A)$, $\mathrm{Ext}(f) \cap \partial_e S(B(H))$ consists of exactly one element.

We may also drop the requirement that the unique extension must be pure. Then we obtain another property.

Definition 3.12 Let H be a separable Hilbert space and A an abelian unital C^*-subalgebra of $B(H)$. We say that A has the **second Kadison-Singer property** if for every $f \in \partial_e S(A)$, $\mathrm{Ext}(f)$ consists of exactly one element.

A priori, it is unclear whether the first Kadison-Singer propery implies the second, since $\mathrm{Ext}(f)$ might contain more elements than $\mathrm{Ext}(f) \cap \partial_e S(B(H))$. Likewise, the one element in $\mathrm{Ext}(f)$ might not be in $\partial_e S(B(H))$, whence the second Kadison-Singer property might not imply the first. However, it turns out that the first and second Kadison-Singer property are in fact equivalent. To prove this, we first need a lemma and note that for every $f \in S(A)$, $\mathrm{Ext}(f)$ is a convex set, whence we can consider its boundary.

Lemma 3.13 *Let H be a separable Hilbert space and A an abelian unital C^*-subalgebra of $B(H)$. For every $f \in \partial_e S(A)$ we have the following identity:*

$$\partial_e \mathrm{Ext}(f) = \mathrm{Ext}(f) \cap \partial_e S(B(H)).$$

Proof $\subseteq$: It is clear that $\partial_e \mathrm{Ext}(f) \subseteq \mathrm{Ext}(f)$. To see that $\partial_e \mathrm{Ext}(f) \subseteq \partial_e S(B(H))$, suppose that $g \in \partial_e \mathrm{Ext}(f)$, that $h_1, h_2 \in S(B(H))$ and that $t \in (0, 1)$ such that $g = t h_1 + (1 - t) h_2$.

Let k_1 and k_2 be the restrictions of h_1 and h_2 to A, respectively. Then, clearly, k_1 and k_2 are both states on A and we have $f = t k_1 + (1 - t) k_2$. Since f is a pure state on A, this means that $k_1 = k_2 = f$.

Therefore, $h_1, h_2 \in \mathrm{Ext}(f)$, and since $g \in \partial_e \mathrm{Ext}(f)$, this means that $g = h_1 = h_2$. Therefore $g \in \partial_e S(B(H))$. Hence $\partial_e \mathrm{Ext}(f) \subseteq \mathrm{Ext}(f) \cap \partial_e S(B(H))$.

$\supseteq$: Suppose that $g \in \mathrm{Ext}(f) \cap \partial_e S(B(H))$ and $t \in (0, 1)$ and $h_1, h_2 \in \mathrm{Ext}(f)$ such that $g = t h_1 + (1 - t) h_2$. Then also $h_1, h_2 \in S(B(H))$ and since $g \in \partial_e S(B(H))$ we then have $h_1 = h_2 = g$. Therefore $g \in \partial_e \mathrm{Ext}(f)$. $\qquad\square$

Theorem 3.14 *Let H be a Hilbert space and A an abelian unital C^*-subalgebra of $B(H)$. Then A has the first Kadison-Singer property if and only if it has the second Kadison-Singer property.*

Proof Suppose A has the first Kadison-Singer property and let $f \in \partial_e S(A)$. Then, by assumption $\text{Ext}(f) \cap \partial_e S(B(H))$ consists of exactly one element, so by Lemma 3.13, $\partial_e \text{Ext}(f)$ consists of exactly one element.

Now, note that $\text{Ext}(f)$ is convex and is a closed subset of the compact set $S(B(H))$. Therefore, $\text{Ext}(f)$ is convex and compact and the Krein-Milman Theorem (B.4) can be applied to it, i.e. $\text{Ext}(f) = \overline{\text{co}(\partial_e \text{Ext}(f))}$. However, $\partial_e \text{Ext}(f)$ consists of exactly one element, whence $\text{co}(\partial_e \text{Ext}(f))$ consists of exactly one element. Therefore, $\text{Ext}(f)$ contains exactly one element, and A has the second Kadison-Singer property.

For the converse, suppose that A has the second Kadison-Singer property and let $f \in \partial_e S(A)$. Then $\text{Ext}(f)$ contains exactly one element, so $\partial_e \text{Ext}(f) = \text{Ext}(f)$ and hence $\partial_e \text{Ext}(f)$ consists of one element as well. By Lemma 3.13, we then know that $\text{Ext}(f) \cap \partial_e S(B(H))$ consists of one element, i.e. A has the first Kadison-Singer property. $\qquad\square$

By the above theorem, we can drop the adjectives 'first' and 'second' and just speak of one property.

Definition 3.15 Let H be a Hilbert space and A an abelian unital C^*-subalgebra of $B(H)$. Then we say that A has the **Kadison-Singer property** if it has either (and hence both) the first or second Kadison-Singer property.

From now on, the main goal of this text is to classify the examples of a Hilbert space H and an abelian unital C^*-subalgebra $A \subseteq B(H)$ that have the Kadison-Singer property.

3.4 Properties of Extensions and Restrictions

The Kadison-Singer property concerns two parts; existence and uniqueness. The following theorem shows that the first is never an issue.

Theorem 3.16 *Let H be a Hilbert space and A a unital abelian C^*-subalgebra of $B(H)$. Furthermore, let $f \in S(A)$. Then $\text{Ext}(f) \neq \emptyset$.*

Proof $f \in S(A)$, so by Proposition 3.5 $\|f\| = 1$. Since $A \subseteq B(H)$ is a linear subspace, there is a functional $g : B(H) \to \mathbb{C}$ that is an extension of f and $\|g\| = \|f\| = 1$, by the Hahn-Banach theorem (see Theorem B.2).

Since $1 \in A \subseteq B(H)$, $g(1) = f(1) = 1$. Using Proposition 3.5 in the reverse direction, it follows that $g \in S(B(H))$. Therefore, $g \in \text{Ext}(f)$, i.e. $\text{Ext}(f) \neq \emptyset$. $\qquad\square$

Now that we know that an extension always exists, we only have to focus on uniqueness when we want to answer the question whether a given algebra has the Kadison-Singer property. By the following proposition, we know more about an extension in the case it is unique. For this, we use the notion of state-like functionals, which is introduced in Definition C.9.

Proposition 3.17 *Suppose H is a Hilbert space and suppose that $A \subseteq B(H)$ is a unital abelian C^*-subalgebra. Furthermore, let $f \in \partial_e S(A)$ such that $\mathrm{Ext}(f) = \{g\}$. Then for each self-adjoint $a \in B(H)$,*

$$g(a) = \sup\{f(b) : b \in A, b \le a\}.$$

Proof By positivity of f, it is clear that the equation holds for any $a \in A$.

Next, suppose $a \notin A$. Then note that $A + \mathbb{C}a$ is a self-adjoint linear subspace of $B(H)$ that contains the unit. Then define

$$\alpha = \sup\{f(b) : b \in A, b \le a\},$$

and, using this, define $h : A + \mathbb{C}a \to \mathbb{C}$, by

$$x + \lambda a \to f(x) + \lambda \alpha.$$

Note that this is well defined, since $x + \lambda a = y + \mu a$ for some $x, y \in A$ and $\lambda, \mu \in \mathbb{C}$ implies $(\mu - \lambda)a = x - y \in A$, so $\mu - \lambda = 0$, since $a \notin A$. Therefore, $\mu = \lambda$ and $x = y$.

Now, h is obviously linear and h also preserves adjoints, since f is a state. Next, we want to show that h is positive on the positive elements of $A + \mathbb{C}a$. So, suppose that $x + \lambda a \ge 0$.

If $\lambda = 0$, then $x \ge 0$, so $h(x + \lambda a) = f(x) \ge 0$. If $\lambda > 0$, then $x \ge -\lambda a$, so $-\lambda^{-1}x \le a$ and $-\lambda^{-1}x \in A$, so $f(-\lambda^{-1}x) \le \alpha$. Therefore,

$$h(x + \lambda a) = f(x) + \lambda \alpha = \lambda(\alpha - f(-\lambda^{-1}x)) \ge 0.$$

Finally, if $\lambda < 0$, then $-\lambda^{-1}x \ge a$, so $f(-\lambda^{-1}x) \ge f(b)$ for every $b \in A$ such that $b \le a \le -\lambda^{-1}x$. Therefore, $f(-\lambda^{-1}x) \ge \alpha$. So certainly,

$$h(x + \lambda a) = f(x) + \lambda \alpha = -\lambda(f(-\lambda^{-1}x) - \alpha) \ge 0.$$

Therefore, h is positive on the positive elements of $A + \mathbb{C}a$, i.e. h is a state-like functional (see Definition C.9). Therefore, by Theorem C.10, h extends to a state-like functional k on $B(H)$. However, state-like functionals on a C^*-algebra are clearly states, so $k \in S(B(H))$. Furthermore, for $x \in A \subseteq A + \mathbb{C}a$, $k(x) = h(x) = f(x)$, i.e. $k \in \mathrm{Ext}(f) = \{g\}$. Therefore, since $a \in A + \mathbb{C}a$, $g(a) = k(a) = h(a) = \alpha$. $\square$

In studying extensions of pure states, it is also useful to understand the reverse direction: restriction. For this, we have the following lemma.

Lemma 3.18 *Suppose A is a C^*-algebra and $C \subseteq A$ a C^*-subalgebra. Then the restriction map*

$$\Phi : S(A) \to S(C), f \mapsto f|_C,$$

is continuous.

Proof Note that the state spaces $S(A)$ and $S(C)$ are endowed with the weak*-topology (see Sect. B.1). Therefore, let $f \in S(C)$, $c \in C$ and $\varepsilon > 0$, i.e. let $B(f, c, \varepsilon) \subseteq S(C)$ be an arbitrary subbase element. We now prove that the set $\Phi^{-1}(B(f, c, \varepsilon)) \subseteq S(A)$ is open.

To do this, let $g \in \Phi^{-1}(B(f, c, \varepsilon))$. Then $|\Phi(g)(c) - f(c)| < \varepsilon$, so there is a $\delta > 0$ such that $|\Phi(g)(c) - f(c)| < \varepsilon - \delta$. Then let $h \in B(g, c, \delta)$. Then

$$
\begin{aligned}
\Phi(h)(c) - f(c)| &\leq |\Phi(h)(c) - \Phi(g)(c)| + |\Phi(g)(c) - f(c)| \\
&< |h(c) - g(c)| + \varepsilon - \delta \\
&< \delta + \varepsilon - \delta \\
&= \varepsilon,
\end{aligned}
$$

whence $h \in \Phi^{-1}(B(f, c, \varepsilon))$. Therefore, $B(g, c, \delta) \subseteq \Phi^{-1}(B(f, c, \varepsilon))$, i.e. the set $\Phi^{-1}(B(f, c, \varepsilon))$ is open. Hence Φ is continuous. $\square$

Chapter 4
Maximal Abelian C*-Subalgebras

In Chap. 3 we introduced the Kadison-Singer property and declared our main goal to be classifying Hilbert spaces H and abelian unital C*-subalgebras $A \subseteq B(H)$ that have this property. In this chapter we show that in order to satisfy the Kadison-Singer property, the subalgebra A needs to be maximal. Next, we will discuss some important examples of such maximal abelian C*-subalgebras.

4.1 Maximal Abelian C*-Subalgebras

For a fixed Hilbert space H, we can consider all unital abelian C*-subalgebras of $B(H)$ and collect them in $C(B(H))$. For every element of $A \in C(B(H))$, we can ask ourselves whether A has the Kadison-Singer property with respect to $B(H)$. It turns out that only maximal elements of $C(B(H))$ can possibly have the Kadison-Singer property with respect to the canonical partial order $\leq$ on $C(B(H))$ given by inclusion, i.e. for $A_1, A_2 \in C(B(H))$ we have $A_1 \leq A_2$ iff $A_1 \subseteq A_2$. Since it would only be tedious to use the symbol $\leq$, we just use the inclusion symbol $\subseteq$ to denote the partial order.

Since $(C(B(H)), \subseteq)$ is now a partially ordered set, we can consider its maximal elements.

Definition 4.1 Suppose H is an Hilbert space and $A_1 \in C(B(H))$. Then A_1 is called **maximal abelian** if it is maximal with respect to the partial order '$\subseteq$' on $C(B(H))$, i.e. if $A_1 \subseteq A_2$ for some $A_2 \in C(B(H))$, then necessarily $A_1 = A_2$.

Maximal abelian elements of $C(B(H))$ have a very nice description in terms of the commutant.

Definition 4.2 Suppose X is an algebra and $S \subseteq X$ is a subset. We define the **commutant** of S to be

© The Author(s) 2016
M. Stevens, *The Kadison-Singer Property*,
SpringerBriefs in Mathematical Physics 14, DOI 10.1007/978-3-319-47702-2_4

$$S' := \{x \in X \mid sx = xs \ \forall s \in S\},$$

i.e. the set of all $x \in X$ that commute with all of S.

We denote the **double commutant** of a subset S of an algebra X by $S'' := (S')'$ and likewise $S''' = (S'')'$. The proofs of the following properties of the commutant are trivial.

Lemma 4.3 *Suppose X is an algebra and $S, T \subseteq X$ are subsets. Then:*

1. $S \subseteq S'$ *iff S is abelian.*
2. *If $S \subseteq T$, then $T' \subseteq S'$.*
3. $S \subseteq S''$.
4. $S' = S'''$.

We can now give a description of maximal abelian subalgebras in terms of the commutant.

Proposition 4.4 *Suppose A is a subalgebra of $B(H)$, for some Hilbert space H. Then the following are equivalent:*

1. $A \in C(B(H))$ *and A is maximal abelian;*
2. $A = A'$.

Proof Suppose $A \in C(B(H))$ is maximal abelian. Since A is abelian, $A \subseteq A'$.

Now let $b \in A'$ and let C be the smallest C*-subalgebra of $B(H)$ that contains A and b. Then since b commutes with all of A, C is abelian and unital, since we have $1 \in A \subseteq C$. Therefore, $C \in C(B(H))$ and $A \subseteq C$. However, A was assumed to be maximal, whence $C = A$. Hence $b \in C = A$ and $A' \subseteq A$, so $A' = A$.

For the converse, suppose that $A = A'$. First note that $1 \in A' = A$ and $A \subseteq A'$, so $A \in C(B(H))$. Now suppose that $C \in C(B(H))$ such that $A \subseteq C$. Then C is abelian, so $C \subseteq C' \subseteq A' = A$, whence $A = C$ and A is maximal. $\square$

The above proposition justifies dropping the adjective 'unital' when we defined maximal abelian subalgebras.

We now come to the main result in this chapter: only maximal abelian subalgebras can possibly have the Kadison-Singer property.

Theorem 4.5 *Suppose that H is a Hilbert space and that $A \in C(B(H))$ has the Kadison-Singer property. Then A is maximal abelian.*

Proof Suppose $C \in C(B(H))$ such that $A \subseteq C$. We will show that the pure state spaces $\partial_e S(C)$ and $\partial_e S(A)$ are isomorphic. To do this, first construct the map:

$$\Phi : \partial_e S(C) \to \partial_e S(A), \ f \mapsto f|_A$$

Since the pure states are exactly the characters on an abelian C*-subalgebra (see Theorem 3.8) and $f|_A$ is therefore a non-zero restriction of a character, we know that $f|_A \in \Omega(A) = \partial_e S(A)$ for all $f \in \partial_e S(C)$. Therefore, Φ is well defined.

For any $g \in \partial_e S(A)$, we know that $\mathrm{Ext}(g)$ contains exactly one element. Denote this element by $\widetilde{g}$. Using this, we can construct the following map:

$$\Psi : \partial_e S(A) \to \partial_e S(C), \ g \mapsto \widetilde{g}|_C$$

To show that this map is well defined, let $g \in \partial_e S(A)$. Note that $\widetilde{g}$ is a state on $B(H)$, and $\widetilde{g}|_C$ is therefore a state on C, since positivity and unitality are clearly preserved under restriction. Now write $h = \widetilde{g}|_C$ and suppose $h = th_1 + (1-t)h_2$ for some $t \in (0,1)$ and $h_1, h_2 \in S(C)$. By Theorem 3.16 we can find $k_1 \in \mathrm{Ext}(h_1)$ and $k_2 \in \mathrm{Ext}(h_2)$. Then $k_1|_A = h_1|_A$ and $k_2|_A = h_2|_A$, so

$$g = \widetilde{g}|_A = h|_A = th_1|_A + (1-t)h_2|_A = tk_1|_A + (1-t)k_2|_A.$$

However, $g \in \partial_e S(A)$, so $k_1|_A = k_2|_A = g$, i.e. $k_1, k_2 \in \mathrm{Ext}(g)$. So $k_1 = k_2 = \widetilde{g}$. Then $h_1 = k_1|_C = \widetilde{g}|_C = h$ and likewise $h_2 = h$, i.e. $h \in \partial_e S(C)$, as desired.

The only thing left to show is that Φ and Ψ are each other's inverse. First, let $g \in \partial_e S(A)$. Then $(\Phi \circ \Psi)(g) = \widetilde{g}|_A = g$, since $\widetilde{g} \in \mathrm{Ext}(g)$. Hence $\Phi \circ \Psi = \mathrm{Id}$.

Next, let $f \in \partial_e S(C)$. Choose $h \in \mathrm{Ext}(f)$, which exists by Theorem 3.16. Then certainly $h \in \mathrm{Ext}(f|_A)$. However, by assumption $\mathrm{Ext}(f|_A)$ contains exactly one element, so $h = \widetilde{f|_A}$. Hence

$$(\Psi \circ \Phi)(f) = \widetilde{(f|_A)}|_C = h|_C = f,$$

since $h \in \mathrm{Ext}(f)$. Therefore, $\Psi \circ \Phi = \mathrm{Id}$.

Hence $\Phi : \partial_e S(C) \to \partial_e S(A)$ is a bijection. It is also continuous by Lemma 3.18. By Corollary 3.9 we know that $\partial_e S(C)$ and $\partial_e S(A)$ are both compact Hausdorff, so by Lemma A.13 Φ is in fact a homeomorphism. Therefore, Φ induces an isomorphism

$$\Phi^* : C_0(\partial_e S(A)) \to C_0(\partial_e S(C))$$

given by $\Phi^*(F)(f) = F(\Phi(f))$.

Using the Gelfand representation (Theorem B.25) twice, i.e. using the isomorphisms

$$G_A : A \to C_0(\Omega(A)) = C_0(\partial_e S(A)), (G_A(a))(f) = f(a)$$

and

$$G_C : C \to C_0(\Omega(C)) = C_0(\partial_e S(C)), (G_C(c))(f) = f(c),$$

we can construct an isomorphism $F = G_C^{-1} \circ \Phi^* \circ G_A$ such that the following diagram commutes:

$$A \xrightarrow{\ G_A\ } C_0(\partial_e S(A))$$

$$\downarrow F \qquad\qquad\qquad \downarrow \Phi^*$$

$$C \xrightarrow{\ G_C\ } C_0(\partial_e S(C))$$

We now claim that F is in fact given by the inclusion map $i : A \to C$. To see this, let $a \in A$ and $f \in \partial_e S(C)$. Then:

$$((\Phi^* \circ G_A)(a))(f) = \Phi^*(G_A(a))(f) = G_A(a)(\Phi(f))$$
$$= \Phi(f)(a) = f|_A(a) = (f \circ i)(a)$$
$$= f(i(a)) = G_C(i(a))(f)$$
$$= ((G_c \circ i)(a))(f).$$

Hence $\Phi^* \circ G_A = G_C \circ i$, so indeed $i = G_C^{-1} \circ \Phi^* \circ G_A = F$. So the inclusion map $i : A \to C$ is an isomorphism, i.e. $A = C$.

Therefore, A is maximal abelian.   □

Thus, in our search for a classification of subalgebras with the Kadison-Singer property, we now merely have to focus on maximal abelian subalgebras.

4.2   Examples of Maximal Abelian C*-Subalgebras

It is time to give some key examples of maximal abelian C*-subalgebras, since these are the only ones that can possess the Kadison-Singer property. In Chap. 2 we proved that $D \subseteq M$ has the Kadison-Singer property (Theorem 2.14). Together with Theorem 4.5, this implies that $D \subseteq M$ is maximal abelian. However, one can also prove this directly by an easy proof.

For infinite-dimensional separable Hilbert spaces, examples of maximal abelian C*-subalgebras become more involved.

4.2.1   The Discrete Subalgebra

One of the most important examples of a Hilbert space is the space $\ell^2(\mathbb{N})$, defined as

$$\ell^2(\mathbb{N}) = \{f : \mathbb{N} \to \mathbb{C} \mid \sum_{n \in \mathbb{N}} |f(n)|^2 < \infty\}.$$

This space has a natural inner product

$$\langle f, g \rangle = \sum_{n \in \mathbb{N}} \overline{f(n)} g(n),$$

which makes $\ell^2(\mathbb{N})$ a Hilbert space. $\ell^2(\mathbb{N})$ is separable because the functions $\{\delta_n\}_{n\in\mathbb{N}}$ defined by $\delta_n(m) = \delta_{nm}$ form a countable basis.

We can also consider the bounded functions on $\mathbb{N}$, given by

$$\ell^\infty(\mathbb{N}) = \{f : \mathbb{N} \to \mathbb{C} \mid \sup_{n\in\mathbb{N}} |f(n)| < \infty\}.$$

It is clear that $\ell^\infty(\mathbb{N})$ is an abelian algebra under pointwise operations. Defining the adjoint operation pointwise as $f^*(n) = \overline{f(n)}$, $\ell^\infty(\mathbb{N})$ becomes a C^*-algebra in the norm

$$\|f\|_\infty = \sup_{n\in\mathbb{N}} |f(n)|.$$

We can now define the very important map

$$M : \ell^\infty(\mathbb{N}) \to B(\ell^2(\mathbb{N})), f \mapsto M_f,$$

defined by

$$(M_f(\phi))(n) = f(n)\phi(n).$$

This is a well-defined norm-preserving injective *-homomorphism, and is called the *multiplication operator*. The proof of this is rather tedious, but mostly trivial.

Because of this fact, we can identify $\ell^\infty(\mathbb{N})$ with the subalgebra $M(\ell^\infty(\mathbb{N}))$ of $B(\ell^2(\mathbb{N}))$. We will tacitly use this identification.

Proposition 4.6 *The subalgebra $\ell^\infty(\mathbb{N}) \subseteq B(\ell^2(\mathbb{N}))$ is maximal abelian.*

Proof $\ell^\infty(\mathbb{N})$ is abelian, so $\ell^\infty(\mathbb{N}) \subseteq \ell^\infty(\mathbb{N})'$.

Now let $T \in \ell^\infty(\mathbb{N})'$. Define $f : \mathbb{N} \to \mathbb{C}$ by

$$f(n) := (T(\delta_n))(n).$$

For every $n \in \mathbb{N}$, $\|\delta_n\| = 1$, so

$$|f(n)|^2 = |(T(\delta_n))(n)|^2 \leq \sum_{m\in\mathbb{N}} |(T(\delta_n))(m)|^2 = \|T(\delta_n)\|^2 \leq \|T\|^2.$$

Therefore, $\sup_{n\in\mathbb{N}} |f(n)| \leq \|T\|$, i.e., $f \in \ell^\infty(\mathbb{N})$.

Now take $\phi \in \ell^2(\mathbb{N})$. Then for any $n, m \in \mathbb{N}$ we have:

$$(M_{\delta_n}(\phi))(m) = \delta_{nm}\phi(m) = \phi(n)\delta_{nm} = \phi(n)\delta_n(m),$$

i.e. $M_{\delta_n}(\phi) = \phi(n)\delta_n$ for all $n \in \mathbb{N}$.

Therefore, for all $n \in \mathbb{N}$:

$$(T(\phi(n))) = ((M_{\delta_n}T)(\phi))(n) = ((TM_{\delta_n})(\phi))(n)$$
$$= \phi(n)(T(\delta_n))(n) = \phi(n)f(n) = (M_f(\phi))(n),$$

where we used the fact that $T \in \ell^\infty(\mathbb{N})'$ and hence commutes with M_{δ_n}.

So, $T(\phi) = M_f(\phi)$, but $\phi \in \ell^2(\mathbb{N})$ was arbitrary, so $T = M_f \in \ell^\infty(\mathbb{N})$. This proves that $\ell^\infty(\mathbb{N})' \subseteq \ell^\infty(\mathbb{N})$. Therefore $\ell^\infty(\mathbb{N}) = \ell^\infty(\mathbb{N})'$, so $\ell^\infty(\mathbb{N}) \subseteq B(\ell^2(\mathbb{N}))$ is maximal abelian. $\square$

There is considerable similarity between the case $D \subseteq M$ that we treated in Chap. 2 and $\ell^\infty(\mathbb{N}) \subseteq B(\ell^2(\mathbb{N}))$; the latter can be viewed as the infinite-dimensional version of the first. We can make this observation more precise by rewriting the case $D \subseteq M$ in a suitable fashion.

To do this, for every $n \in \mathbb{N}$ write $\underline{n} = \{1, \ldots, n\}$ and define

$$\ell(\underline{n}) = \{f : \underline{n} \to \mathbb{C}\}.$$

Note that in comparison with the infinite case, in this case it does not matter whether we take all functions (like we did now), or the square-summable functions (which would give $\ell^2(\underline{n})$) or the bounded functions ($\ell^\infty(\underline{n})$), since these are all the same.

Furthermore, we can endow $\ell(\underline{n})$ with a canonical inner product

$$\langle f, g \rangle = \sum_{k \in \underline{n}} \overline{f(k)} g(k)$$

which makes $\ell(\underline{n})$ a Hilbert space. As a Hilbert space, $\ell(\underline{n})$ is clearly isomorphic to $\mathbb{C}^n$ under the canonical isomorphism

$$\ell(\underline{n}) \to \mathbb{C}^n, \ f \mapsto (f(1), \ldots, f(n)).$$

This isomorphism induces an isomorphism between operators on $\ell(\underline{n})$ and operators on $\mathbb{C}^n$, explicitly given by

$$\varphi : B(\ell(\underline{n})) \to M_n(\mathbb{C}), \ \varphi(T)_{ij} = (T(\delta_j))(i).$$

Just as in the infinite-dimensional case, we can define a multiplication operator

$$M : \ell(\underline{n}) \to B(\ell(\underline{n})), f \mapsto M_f, M_f(\phi)(m) = f(m)\phi(m)$$

Since we are now dealing with the finite case, there is no question whether this map is well defined, since all linear operators are automatically bounded. Just like in the infinite case, we can identify $\ell(\underline{n})$ with $M(\ell(\underline{n})) \subseteq B(\ell(\underline{n}))$.

The main point is the following: it is easy to see that the diagonal matrices, as discussed in Chap. 2, exactly correspond to the multiplication operators. To be more precise, we have that for every $n \in \mathbb{N}$ the restriction of the isomorphism $\varphi : B(\ell(\underline{n})) \to M_n(\mathbb{C})$ to $\ell(\underline{n})$ gives an isomorphism between $\ell(\underline{n})$ and $D_n(\mathbb{C})$.

Summarizing, we see that the finite-dimensional case and the infinite-dimensional case are not that different. Therefore, we introduce one general description. Let $\aleph_0$ denote the cardinality of $\mathbb{N}$ and write $\aleph_0 = \mathbb{N}$. The expression '$1 \leq j \leq \aleph_0$' means 'either $j \in \mathbb{N}$ or $j = \aleph_0$'. This can be made more precise by adding a maximal element $\aleph_0$ to the totally ordered set $\mathbb{N}$.

Definition 4.7 Let $1 \leq j \leq \aleph_0$. Then $A_d(j)$ is the subalgebra $\ell^\infty(\underline{j}) \subseteq B(\ell^2(\underline{j}))$ that acts on the Hilbert space $\ell^2(\underline{j})$ via the multiplication operator. We call $A_d(j)$ the **discrete subalgebra of cardinality** j.

Note that we have used the identification $\ell(\underline{j}) = \ell^2(\underline{j}) = \ell^\infty(\underline{j})$ for $j \in \mathbb{N}$. Discrete subalgebras provide key examples of maximal abelian subalgebras and will play a major role in our further discussion.

4.2.2 The Continuous Subalgebra

Another important example of a maximal abelian subalgebra is non-discrete. As an introduction to this example, we consider all measurable functions from $[0, 1]$ to $\mathbb{C}$:

$$\mathscr{F}(0, 1) := \{ f : [0, 1] \to \mathbb{C} \mid f \text{ is measurable} \},$$

where we use the standard Lebesgue measure μ on $[0, 1]$. We define a relation $\sim$ on $\mathscr{F}(0, 1)$ by

$$f \sim g \iff \mu(\{x \in [0, 1] : f(x) \neq g(x)\}) = 0.$$

We sometimes denote the latter condition as $\mu(f \neq g) = 0$. It is clear that $\sim$ is an equivalence relation on $\mathscr{F}(0, 1)$, so we can define:

$$F(0, 1) := \mathscr{F}(0, 1)/ \sim .$$

We denote equivalence classes in $F(0, 1)$ by $[f]$, where $f \in \mathscr{F}(0, 1)$ is a representative. $F(0, 1)$ is an algebra under the canonical operations $\lambda[f] + [g] = [\lambda f + g]$ and $[f][g] = [fg]$. Using this, it is easy to see that the function

$$I_2 : F(0, 1) \to [0, \infty], \quad [f] \mapsto \int_{[0,1]} |f(x)|^2 \mathrm{d}x$$

is well defined.

Then, we can define a new space, which we call the **space of square-integrable functions**:

$$L^2(0, 1) := \{ \psi \in F(0, 1) \mid I_2(\psi) < \infty \}.$$

One of the most important results of basic functional analysis is that $L^2(0, 1)$ is a Hilbert space with respect to the inner product $\langle\,,\,\rangle$, given by:

$$\langle[f], [g]\rangle = \int_{[0,1]} \overline{f(x)}g(x)\mathrm{d}x.$$

The equivalence relation $\sim$ is necessary in the construction of $L^2(0, 1)$ in order for the inner product on $L^2(0, 1)$ to be positive definite. Note that the norm induced by this inner product satisfies $\|\psi\|^2 = I_2(\psi)$.

There is a certain kind of analogy between $L^2(0, 1)$ and $\ell^2(\mathbb{N})$, by replacing sums by integrals. Just as in the case of $\ell^2(\mathbb{N})$ one could again want to define the space of bounded functions. Because we are dealing with equivalence classes of functions, we need to define this properly: we put

$$L^\infty(0, 1) := \{\psi \in F(0, 1) \mid \exists f \in \psi : \sup_{x\in[0,1]} |f(x)| < \infty\}.$$

This is called the **space of essentially bounded functions**, coming with a natural norm:

$$\|\psi\|_\infty^{(\mathrm{ess})} = \inf_{f\in\psi}\{k \in [0, \infty) : |f(x)| \le k \,\forall x \in [0, 1]\}.$$

If we include the operation $[f]^* = [\overline{f}]$, then $L^\infty(0, 1)$ becomes a C*-algebra.

Now we have made our set-up: similar to the previous example, we want to regard $L^\infty(0, 1)$ as a subalgebra of $B(L^2(0, 1))$. Again, we do this by means of a **multiplication operator**:

$$M : L^\infty(0, 1) \to B(L^2(0, 1)), \ \psi \mapsto M_\psi,$$

where $M_{[f]}([g]) = [fg]$.

Just as in the discrete case, it can be shown that M is a well-defined injective, norm-preserving, *-homomorphism. Therefore, we can regard $L^\infty(0, 1)$ as a C*-subalgebra of $B(L^2(0, 1))$, where we tacitly identify $L^\infty(0, 1)$ with its image under M. Of course, $L^\infty(0, 1)$ is an abelian subalgebra. We introduced this example since it is *maximal* abelian.

Theorem 4.8 $L^\infty(0, 1) \subseteq B(L^2(0, 1))$ *is maximal abelian.*

Proof $L^\infty(0, 1)$ is abelian, so $L^\infty(0, 1) \subseteq L^\infty(0, 1)'$.

For the other inclusion, suppose that $T \in L^\infty(0, 1)'$. Note that $I_2([1]) = 1$, so $[1] \in L^2(0, 1)$. Therefore, we can define $\psi = T([1]) \in L^2(0, 1)$. We claim that $\psi \in L^\infty(0, 1)$.

To see this, we argue by contraposition, so we suppose that $\psi \notin L^\infty(0, 1)$. Now let $f \in \psi$ and for every $N \in \mathbb{N}$, define:

$$A_N := \{x \in [0, 1] \ : \ |f(x)| \ge N\}.$$

Since $\psi \notin L^{\infty}(0, 1)$, for every $N \in \mathbb{N}$, $\mu(A_N) \neq 0$. Since $1_{A_N} \in L^{\infty}(0, 1)$, we can compute:

$$T([1_{A_N}]) = T(M_{[1_{A_N}]}([1])) = M_{[1_{A_N}]}(T([1])) = M_{[1_{A_N}]}([f]) = [f \cdot 1_{A_N}].$$

Therefore, we also have:

$$N^2 \mu(A_N) \leq \int_{A_N} |f(x)|^2 dx = \|[f \cdot 1_{A_N}]\|^2 = \|T([1_{A_N}])\|^2$$
$$\leq \|T\|^2 \|[1_{A_N}]\|^2 = \|T\|^2 \mu(A_N).$$

Since $\mu(A_N) \neq 0$, $N \leq \|T\|$ for all $N \in \mathbb{N}$. However, $T \in B(L^2(0, 1))$, so this is a contradiction. Hence $\psi \in L^{\infty}(0, 1)$.

We now claim that $T = M_{\psi}$. To see this, let $\phi \in L^2(0, 1)$ and let $g \in \phi$. For each $n \in \mathbb{N}$ define

$$U_n := \{x \in [0, 1] : |g(x)| \leq n\},$$

and $g_n := g \cdot 1_{U_n}$. Note that the sequence of functions $f_i : [0, 1] \to [0, \infty)$ defined by $f_i(x) = |g_i(x)|^2$ is pointwise non-decreasing and has $f : [0, 1] \to [0, \infty)$, $f(x) = |g(x)|^2$, as its pointwise limit. Hence, by Lebesgue's monotone convergence theorem,

$$\lim_{n \to \infty} \|[g_n]\|^2 = \lim_{n \to \infty} \int_{[0,1]} |g_n(x)|^2 \, dx = \int_{[0,1]} |g(x)|^2 \, dx = \|[g]\|^2.$$

Furthermore,

$$\|[g] - [g_n]\|^2 = \int_{[0,1] \setminus U_n} |g(x)|^2 \, dx = \int_{[0,1]} |g(x)|^2 \, dx - \int_{U_n} |g(x)|^2 \, dx$$
$$= \|[g]\|^2 - \|[g_n]\|^2,$$

whence $\lim_{n \to \infty} \|[g] - [g_n]\| = 0$, i.e. $\lim_{n \to \infty} [g_n] = [g]$.

Choose $h \in \psi$. Since $[g_n] \in L^{\infty}(0, 1)$, we can compute:

$$T([g_n]) = T(M_{[g_n]}([1])) = M_{[g_n]}(T([1]))$$
$$= M_{[g_n]}([h]) = [g_n h]$$
$$= M_{[h]}([g_n]) = M_{\psi}([g_n]).$$

Then also, by continuity of both T and M_{ψ},

$$T([g]) = T(\lim_{n \to \infty} [g_n]) = \lim_{n \to \infty} T([g_n]) = \lim_{n \to \infty} M_{\psi}([g_n]) = M_{\psi}(\lim_{n \to \infty} [g_n]) = M_{\psi}([g]).$$

Therefore, $T(\phi) = M_\psi(\phi)$. Since $\phi \in L^2(0, 1)$ was arbitrary, $T = M_\psi$. So, we conclude that $T \in L^\infty(0, 1)$. Hence $L^\infty(0, 1)' \subseteq L^\infty(0, 1)$. Therefore, we know that $L^\infty(0, 1)' = L^\infty(0, 1)$, i.e. $L^\infty(0, 1)$ is maximal abelian.   □

Along the lines of the definition of the discrete subalgebra of cardinality j, we introduce a special short notation for the subalgebra $L^\infty(0, 1) \subseteq B(L^2(0, 1))$.

Definition 4.9 We denote the maximal abelian subalgebra $L^\infty(0, 1)$ of $B(L^2(0, 1))$ by A_c, realized via multiplication operators. We call A_c the **continuous subalgebra**.

4.2.3 The Mixed Subalgebra

Combining two different examples of maximal abelian subalgebras, one can construct another example of a maximal abelian subalgebra. Here, we use the notation as introduced in the appendix, most notably in Sect. B.2.

Proposition 4.10 *Suppose $A_1 \subseteq B(H_1)$ and $A_2 \subseteq B(H_2)$ are both maximal abelian C*-subalgebras. Then $A_1 \oplus A_2 \subseteq B(H_1 \oplus H_2)$ is maximal abelian.*

Proof Since $A_1 \oplus A_2(j)$ is a pointwise defined subalgebra of $B(H_1 \oplus H_2)$ and both A_1 and A_2 are abelian, $A_1 \oplus A_2$ is abelian. Therefore $A_1 \oplus A_2 \subseteq (A_1 \oplus A_2)'$. Next, suppose that $T \in (A_1 \oplus A_2)'$. Define $T_1 = \pi_1 \circ T \circ \iota_1$ and $T_2 = \pi_2 \circ T \circ \iota_2$. Since T is bounded, $T_1 \in B(H_1)$ and $T_2 \in B(H_2)$.

Now note that for any $x \in H_1$ and $y \in H_2$,

$$
\begin{aligned}
T(x, y) &= T(\iota_1(x) + \iota_2(y)) \\
&= T(\iota_1(x)) + T(\iota_2(y)) \\
&= (T \circ (1, 0) \circ \iota_1)(x) + (T \circ (0, 1) \circ \iota_2)(y) \\
&= ((1, 0) \circ T \circ \iota_1)(x) + ((0, 1) \circ T \circ \iota_2)(y) \\
&= ((\pi_1 \circ T \circ \iota_1)(x), 0) + (0, (\pi_2 \circ T \circ \iota_2)(y)) \\
&= (T_1(x), 0) + (0, T_2(y)) \\
&= (T_1(x), T_2(y)),
\end{aligned}
$$

where we used the fact that T commutes with $(1, 0)$ and $(0, 1)$, since $T \in (A_1 \oplus A_2)'$. Therefore, $T = (T_1, T_2)$. Now, for all $a \in A_1$,

$$(T_1 \circ a, 0) = T \circ (a, 0) = (a, 0) \circ T = (a \circ T_1, 0)$$

Therefore, $T_1 \in A_1' = A_1$. Likewise, $T_2 \in A_2$. Hence $T = (T_1, T_2) \in A_1 \oplus A_2$, i.e. $(A_1 \oplus A_2)' \subseteq A_1 \oplus A_2$. Therefore

$$(A_1 \oplus A_2)' = A_1 \oplus A_2,$$

i.e. $A_1 \oplus A_2 \subseteq B(H_1 \oplus H_2)$ is maximal abelian. $\square$

Since we are interested in the question whether a maximal abelian subalgebra possesses the Kadison-Singer property, we would like to make a connection between the Kadison-Singer property for a direct sum $A_1 \oplus A_2$ and the Kadison-Singer property of A_1 and A_2 separately. It turns out that we can do this. First of all, we need to describe the characters (and hence the pure states) of a direct sum. For this, note that for any map $f : A_i \to \mathbb{C}$, the pullback over the projection $\pi_i : A_1 \oplus A_2 \to A_i$, i.e. $\pi_i^*(f) = f \circ \pi_i$, gives a map $\pi_i^*(f) : A_1 \oplus A_2 \to \mathbb{C}$.

Proposition 4.11 *Suppose A_1 and A_2 are both C*-algebras. Then*

$$\Omega(A_1 \oplus A_2) = \pi_1^*(\Omega(A_1)) \cup \pi_2^*(\Omega(A_2)).$$

Proof Suppose $f \in \Omega(A_1 \oplus A_2)$. Then

$$f((0, 1))^2 = f((0, 1)^2) = f((0, 1)),$$

so $f((0, 1)) \in \{0, 1\}$. Likewise $f((1, 0)) \in \{0, 1\}$. However, we also have

$$f((0, 1)) + f((1, 0)) = f((1, 1)) = f(1) = 1,$$

so there are two cases. Either $f((1, 0)) = 1$ and $f((0, 1)) = 0$, or $f((1, 0)) = 0$ and $f((0, 1)) = 1$.

Suppose the first case is true. Then define $g : A_1 \to \mathbb{C}$ by $g(a) = f(a, 0)$. Then $g(1) = 1$, so g is non-zero and for any $a_1, a_2 \in A_1$ we have

$$g(a_1 a_2) = f((a_1 a_2, 0)) = f((a_1, 0)) f((a_2, 0)) = g(a_1) g(a_2),$$

so $g \in \Omega(A_1)$. Furthermore, for any $(a_1, a_2) \in A_1 \oplus A_2$ we have

$$\begin{aligned}
f((a_1, a_2)) &= f((a_1, 0)) + f((0, a_2)) = f((a_1, 0)) f((1, 0)) + f((0, a_2)) \\
&= f((a_1, 0)) = g(a_1) = (g \circ \pi_1)((a_1, a_2)),
\end{aligned}$$

i.e. $f = \pi_1^*(g)$, so $f \in \pi_1^*(\Omega(A_1))$.

If the second case is true, it follows likewise that $f \in \pi_2^*(S(A_2))$. Hence

$$\Omega(A_1 \oplus A_2) \subseteq \pi_1^*(\Omega(A_1)) \cup \pi_2^*(\Omega(A_2)).$$

Now suppose $h \in \pi_1^*(\Omega(A_1))$. Then $h = k \circ \pi_1$ for some $k \in \Omega(A_1)$, so

$$h(1) = h((1, 1)) = k(1) = 1,$$

i.e. h is non-zero. Furthermore, h is clearly linear and for any pair of elements $(a_1, a_2), (b_1, b_2) \in A_1 \oplus A_2$, we have

$$h((a_1, a_2)(b_1, b_2)) = h((a_1b_1, a_2b_2)) = k(a_1b_1)$$
$$= k(a_1)k(b_1) = h((a_1, a_2))h((b_1, b_2)),$$

i.e. $h \in \Omega(A_1 \oplus A_2)$. Therefore, $\pi_1^*(A_1) \subseteq \Omega(A_1 \oplus A_2)$. Likewise, we have that $\pi_2^*(\Omega(A_2)) \subseteq \Omega(A_1 \oplus A_2)$, so indeed, $\Omega(A_1 \oplus A_2) = \pi_1^*(\Omega(A_1)) \cup \pi_2^*(\Omega(A_2))$. □

The above proposition gives us information about the pure states on a direct sum of abelian subalgebras, since the pure states are exactly the characters. Next, we need to make a connection between the concepts of positivity and direct sums of operator algebras.

Lemma 4.12 *Suppose H_1 and H_2 are Hilbert spaces and $b \in B(H_1 \oplus H_2)$ is positive. Then for $j \in \{1, 2\}$, $\pi_j b i_j \in B(H_j)$ is positive.*

Proof Let $(x, y) \in H_1 \oplus H_2$. Then compute:

$$\langle (\pi_1 b i_1)(x), x \rangle = \langle (\pi_1 b)((x, 0), x \rangle$$
$$= \langle (\pi_1 b)((x, 0), x \rangle + \langle (\pi_2 b)(x, 0), 0 \rangle$$
$$= \langle b(x, 0), (x, 0) \rangle \geq 0,$$

since b is positive. Therefore, $\pi_1 b i_1$ is positive. Likewise, $\pi_2 b i_2$ is positive. □

We use these results to prove the following theorem about the connection between direct sums and the Kadison-Singer property.

Theorem 4.13 *Suppose H_1 and H_2 are Hilbert spaces. Furthermore, suppose that $A_1 \subseteq B(H_1)$ and $A_2 \subseteq B(H_2)$ are abelian unital C*-subalgebras such that the subalgebra $A_1 \oplus A_2 \subseteq B(H_1 \oplus H_2)$ has the Kadison-Singer property. Then, both $A_1 \subseteq B(H_1)$ and $A_2 \subseteq B(H_2)$ have the Kadison-Singer property.*

Proof Suppose $f \in \partial_e S(A_1)$ and $g_1, g_2 \in \mathrm{Ext}(f) \subseteq B(H_1)$. Then $f \in \Omega_1$, so by Proposition 4.11, $\pi_1^*(f) \in \Omega(A_1 \oplus A_2) = \partial_e S(A_1 \oplus A_2)$.

Now define the linear functionals $k_1, k_2 : B(H_1 \oplus H_2) \to \mathbb{C}$ by $k_j(b) = g_j(\pi_1 b i_1)$ for all $b \in B(H_1 \oplus H_2)$ and $j \in \{1, 2\}$. Then for $j \in \{1, 2\}$,

$$k_j(1) = g_j(\pi_1 i_1) = g_j(1) = 1,$$

since g_j is a state. Furthermore for a positive $b \in B(H_1 \oplus H_2)$, $\pi_1 b i_1 \in B(H_1)$ is positive by Lemma 4.12. Therefore, $k_j(b) = g_j(\pi_1 b i_1) \geq 0$, since g_j is positive. Hence $k_1, k_2 \in S(B(H_1 \oplus H_2))$.

Now, for an element $(a_1, a_2) \in A_1 \oplus A_2$, $\pi_1(a_1, a_2)i_1 = a_1$, so

$$k_j((a_1, a_2)) = g_j(\pi_1(a_1, a_2)i_1) = g_j(a_1)$$
$$= f(a_1) = (f \circ \pi_1)(a_1, a_2) = \pi_1^*(f)((a_1, a_2)),$$

i.e. $k_1, k_2 \in \text{Ext}(\pi_1^*(f))$. However, by assumption, $A_1 \oplus A_2 \subseteq B(H_1 \oplus H_2)$ has the Kadison-Singer property, so $\text{Ext}(\pi_1^*(f))$ has at most one element, i.e. $k_1 = k_2$.

For any $b \in B(H_1)$, $b = \pi_1(b, 0)i_1$, so we have

$$g_1(b) = g_1(\pi_1(b, 0)i_1) = k_1((b, 0)) = k_2((b, 0)) = g_2((\pi_1(b, 0)i_1) = g_2(b),$$

i.e. $g_1 = g_2$. Therefore, $\text{Ext}(f)$ has at most one element. Combined with Theorem 3.16, we know $\text{Ext}(f)$ has exactly one element. Therefore, $A_1 \subseteq B(H_1)$ has the Kadison-Singer property. Likewise, $A_2 \subseteq B(H_2)$ has the Kadison-Singer property.

$\square$

As a special example of a direct sum, we can combine the discrete subalgebra $A_d(j)$ for some $1 \leq j \leq \aleph_0$ with the continuous example A_c. To do this, define

$$H_j := L^2(0, 1) \oplus \ell^2(\underline{j}).$$

We will call the maximal abelian subalgebra $A_c \oplus A_d(j) \subseteq B(H_j)$ the **mixed subalgebra**. As it will turn out later, this is in some way the only direct sum that we need to consider.

By now, we have constructed three different examples: the discrete, continuous and mixed subalgebra. These are all examples with a *separable* Hilbert space. In our search for examples of maximal abelian subalgebras that satisfy the Kadison-Singer property, we will restrict ourselves to this kind of Hilbert spaces, since it turns out that we can make a complete classification of abelian subalgebras with the Kadison-Singer property when we only consider separable Hilbert spaces.

Chapter 5
Minimal Projections in Maximal Abelian von Neumann Algebras

Recall that we are considering maximal abelian C*-subalgebras of $B(H)$, for some Hilbert space H. Note that a maximal abelian C*-subalgebra $A \subseteq B(H)$ satisfies the equation $A' = A$ and A' is a von Neumann algebra by Proposition B.31. Therefore, every maximal abelian C*-subalgebra is a von Neumann algebra. Furthermore, every von Neumann algebra is a C*-algebra (viz. Proposition B.30), so certainly every maximal abelian von Neumann algebra (i.e. a von Neumann algebra A that satisfies $A' = A$) is a maximal abelian C*-algebra. Hence we see that the maximal abelian von Neumann algebras are exactly the maximal abelian C*-algebras.

We will first show that it is only necessary to classify all maximal abelian subalgebras up to unitary equivalence, in order to determine whether they satisfy the Kadison-Singer property. Next, we restrict ourselves to separable Hilbert spaces and by considering maximal abelian subalgebras to be von Neumann algebras, we can classify these subalgebras up to unitary equivalence, by using the existence and properties of minimal projections. Together, this greatly simplifies the classification of subalgebras with the Kadison-Singer property in the case of separable Hilbert spaces.

5.1 Unitary Equivalence

The classification of maximal abelian von Neumann algebras is up to so-called *unitary equivalence*. For this, we need unitary elements.

Definition 5.1 Suppose H and H' are Hilbert spaces. Then $u \in B(H, H')$ is called **unitary** if for all $x, y \in H$, $\langle ux, ux \rangle = \langle x, y \rangle$ and $u(H) = H'$.

The above conditions for being unitary are not always the easiest to check. However, it is easy to show that $u \in B(H, H')$ is unitary if and only if $u^*u = 1$ and $uu^* = 1$.

Using unitary elements, we can define the notion of unitary equivalence of subalgebras of $B(H)$.

© The Author(s) 2016
M. Stevens, *The Kadison-Singer Property*,
SpringerBriefs in Mathematical Physics 14, DOI 10.1007/978-3-319-47702-2_5

Definition 5.2 Suppose that H_1 and H_2 are Hilbert spaces and that $A_1 \subseteq B(H_1)$, $A_2 \subseteq B(H_2)$ are subalgebras. Then A_1 is called **unitarily equivalent** to A_2 if there is a unitary $u \in B(H_1, H_2)$ such that $uA_1u^* = A_2$. We denote this by $A_1 \cong A_2$.

Of course, it is easily proven that unitarily equivalence is indeed an equivalence relation. One of the crucial steps in this chapter is the following theorem: it shows that we only have to consider subalgebras up to unitary equivalence when determining whether the subalgebra satisfies the Kadison-Singer property.

Theorem 5.3 *Suppose that H_1 and H_2 are Hilbert spaces and that $A_1 \subseteq B(H_1)$ and $A_2 \subseteq B(H_2)$ are unital abelian subalgebras that are unitarily equivalent. Then A_1 has the Kadison-Singer property if and only if A_2 has the Kadison-Singer property.*

Proof Suppose that A_1 has the Kadison-Singer property. By assumption, there is a unitary $u \in B(H_1, H_2)$ such that $uA_1u^* = A_2$.

Now let $f \in \partial_e S(A_2)$. Then define $g : A_1 \to \mathbb{C}$ by $g(a) = f(uau^*)$. We first claim that $g \in S(A_1)$. To see this, first let $a \in A_1$ and observe that

$$g(a^*a) = f(ua^*au^*) = f((au^*)^*au^*) \geq 0,$$

since f is positive. Hence g is positive. Furthermore, $g(1) = f(uu^*) = f(1) = 1$, so g is unital too. Hence, indeed $g \in S(A_1)$.

Next, we prove that in fact $g \in \partial_e S(A_1)$. To see this, suppose that $h_1, h_2 \in S(A_1)$ and $t \in (0, 1)$ such that $g = th_1 + (1-t)h_2$. Now define the functional $k_1 : A_2 \to \mathbb{C}$ by $k_1(a) = h_1(u^*au)$ for all $a \in A_2$ and likewise define the map $k_2 : A_2 \to \mathbb{C}$ by $k_2(a) = h_2(u^*au)$ for all $a \in A_2$. Then by the same reasoning as above, $k_1, k_2 \in S(A_2)$. Furthermore, for $a \in A_2$,

$$\begin{aligned}
f(a) &= f(uu^*auu^*) = g(u^*au) \\
&= th_1(u^*au) + (1-t)h_2(u^*au) \\
&= tk_1(a) + (1-t)k_2(a),
\end{aligned}$$

i.e. $f = tk_1 + (1-t)k_2$. However, $f \in \partial_e S(A_2)$ by assumption, so $f = k_1 = k_2$. Then for $a \in A_1$:

$$h_1(a) = h_1(u^*uau^*u) = k_1(uau^*) = f(uau^*) = g(a),$$

i.e. $h_1 = g$. Likewise, $h_2 = g$, so indeed $g \in \partial_e S(A_1)$.

We want to prove that $\mathrm{Ext}(f)$ contains exactly one element. By Theorem 3.16, we know that $\mathrm{Ext}(f) \neq \emptyset$. Therefore, suppose that $c, d \in \mathrm{Ext}(f) \subseteq S(B(H_2))$. Then define $\tilde{c} : B(H_1) \to \mathbb{C}$ by $\tilde{c}(b) = c(ubu^*)$ and likewise define $\tilde{d} : B(H_1) \to \mathbb{C}$ by $\tilde{d}(b) = d(ubu^*)$. Then by the same reasoning as above, $\tilde{c}, \tilde{d} \in S(B(H_1))$.

Now for $a \in A_1$, $uau^* \in A_2$, so

$$\tilde{c}(a) = c(uau^*) = f(uau^*) = g(a),$$

since $c \in \text{Ext}(f)$. Hence $\tilde{c} \in \text{Ext}(g)$. Likewise, $\tilde{d} \in \text{Ext}(g)$. However, A_1 has the Kadison-Singer property, so $\text{Ext}(g)$ has exactly one element, i.e. $\tilde{c} = \tilde{d}$.

Let $b \in B(H_2)$. Then

$$c(b) = c(uu^*buu^*) = \tilde{c}(u^*bu) = \tilde{d}(u^*bu) = d(uu^*buu^*) = d(b),$$

i.e. $c = d$. Hence $\text{Ext}(f)$ contains exactly one element, so A_2 has the Kadison-Singer property.

Likewise, if A_2 has the Kadison-Singer property, then A_1 has the Kadison-Singer property. $\qquad\qquad\square$

So, using the above theorem, our first main goal is now to classify all maximal abelian subalgebras up to unitary equivalence. We can make this classification when restricting ourselves to separable Hilbert spaces, so we will only consider those from now on.

5.2 Minimal Projections

An important property of a von Neumann algebra is that it is generated by its projections (see Proposition B.32). Considering maximal abelian von Neumann algebras, the set of projections becomes even more important, because it has more structure than in the general case.

To be more precise, write $P(A) = \mathscr{P}(H) \cap A$ for the set of projections in some maximal abelian von Neumann algebra $A \subseteq B(H)$. Since A is abelian, the product of any two elements in $P(A)$ is again an element of $P(A)$ and since A is unital, $P(A)$ is a monoid.

Now write $P_m(A)$ for the set of minimal projections in $P(A)$, where minimal projections are defined as in Definition B.14. The key in the classification of maximal abelian von Neumann algebras lies in the properties of these sets of minimal projections.

As a first step in this classification, we determine $P_m(A)$ for the cases that A is the discrete, continuous, or mixed subalgebra.

Proposition 5.4 Let $1 \le j \le \aleph_0$. Then $P_m(A_d(j)) = \{\delta_n : \underline{j} \to \mathbb{C} : n \in \underline{j}\}$, where $\delta_n(m) = \delta_{nm}$.

Proof Let us first determine the projections in $A_d(j)$. So, suppose that $p \in A_d(j)$ is a projection. Then $p : \underline{j} \to \mathbb{C}$ such that $p^2 = p^* = p$. Then for any $n \in \underline{j}$,

$$p(n)^2 = \overline{p(n)} = p(n),$$

i.e. $p(n) \in \{0, 1\}$. Therefore, $p = 1_A$ for some subset $A \subseteq \underline{j}$.

Since $\sup_{n \in \underline{j}} |1_A(n)| \leq 1$, we also have that $1_A \in A_d(j)$ for every $A \subseteq \underline{j}$. Since it is clear that $1_A^2 = 1_A^* = 1_A$ for every $A \subseteq \underline{j}$, we conclude that the set of projections in $A_d(j)$ is exactly given by $\{1_A : A \subseteq \underline{j}\}$.

Now note that $1_A = 0$ if and only if $A = \emptyset$ and $1_B - 1_A \geq 0$ if and only if $A \subseteq B$. Now suppose $A \subseteq \underline{j}$ is such that 1_A is a minimal projection. Then $A \neq \emptyset$. Suppose $B \subseteq A$. Then $0 \leq 1_B \leq 1_A$, so $1_B = 0$ or $1_B = 1_A$, i.e. $B = \emptyset$ or $B = A$. Hence A consists of exactly one element.

By the same reasoning, for every $A \subseteq \underline{j}$ that has exactly one element, 1_A is a minimal projection. Hence the set of minimal projections in $A_d(j)$ is exactly given by

$$\{1_A : A \subseteq \underline{j}, \#A = 1\} = \{\delta_n : n \in \underline{j}\}. \qquad \square$$

For the discussion of the continuous subalgebra, we first need a few extra ingredients. For any measurable function $f : [0, 1] \to \mathbb{C}$, define

$$U_f = \{x \in [0, 1] : f(x) \notin \{0, 1\}\}.$$

Lemma 5.5 *The map $\chi : A_c \to [0, 1]$ given by $\chi([f]) = \mu(U_f)$ is well defined.*

Proof Since $f : [0, 1] \to \mathbb{C}$ is measurable if $[f] \in A_c$, $U_f \subseteq [0, 1]$ is a measurable set for every $[f] \in A_c$ and hence $\mu(U_f) \in [0, 1]$ is well-defined.

Therefore, the only thing left to check is that the definition of χ is independent of the choice of representative. So, suppose $[f] = [g] \in A_c$.

Then let $C := \{x \in [0, 1] : f(x) \neq g(x)\}$. By assumption, $\mu(C) = 0$. Now suppose $x \notin U_g \cup C$. Then $f(x) = g(x) \in \{0, 1\}$, so $x \notin U_f$. Therefore, we obtain that $U_f \subseteq U_g \cup C$. Then

$$\mu(U_f) \leq \mu(U_g \cup C) \leq \mu(U_g) + \mu(C) = \mu(U_g).$$

By symmetry, we also have $\mu(U_g) \leq \mu(U_f)$, so $\mu(U_f) = \mu(U_g)$ and hence χ is well-defined. $\qquad \square$

We can now characterize the projections in A_c using the map χ.

Lemma 5.6 *Suppose $\psi \in A_c$. Then ψ is a projection if and only if $\chi(\psi) = 0$.*

Proof Suppose $\chi(\psi) \neq 0$. Then for $f \in \psi$, $\mu(U_f) \neq 0$, so

$$\mu(\{x \in [0, 1] : f(x)^2 = f(x)\}) = \mu(\{x \in [0, 1] : f(x) \notin \{0, 1\}\}) = \mu(U_f) \neq 0,$$

whence $[f]^2 = [f^2] \neq [f]$. Therefore, $\psi = [f]$ is not a projection.

Now suppose that $\chi(\psi) = 0$. Again, take an $f \in \psi$. Then $\mu(U_f) = 0$. Now define $h : [0, 1] \to \mathbb{C}$ by $h = f \cdot 1_{[0,1]\setminus U_f}$. Then by construction h is measurable and we have $[h] = [f] = \psi$. Furthermore, $h(x) \in \{0, 1\}$ for every $x \in [0, 1]$, so certainly $h(x)^2 = \overline{h(x)} = h(x)$ for every $x \in \{0, 1\}$. Therefore $[h]^2 = [h] = [h]^*$. Since $\psi = [h]$, ψ is a projection. $\qquad\square$

Using this characterization of projections in A_c, we can prove the following statement.

Proposition 5.7 A_c *has no minimal projections.*

Proof Suppose $\psi \in A_c$ is a non-zero projection. Choose a $f \in \psi$. Then by Lemma 5.6, $\chi(\psi) = 0$, so $\mu(U_f) = 0$. Then define $h = f \cdot 1_{[0,1]\setminus U_f}$ and observe that we then have $[h] = [f] = \psi$.

Since $\psi \neq 0$, $N_h = \{x \in [0, 1] : h(x) \neq 0\}$ has non-zero measure, so there is a $M \subseteq N_h$ such that $0 < \mu(M) < \mu(N_h)$. Now note that $(h - 1_M) \geq 0$ and $[1_M]$ is a projection, whence $[1_M] \leq h$. Furthermore, $\mu(M) \neq 0$, so $[1_M] \neq 0$ and $[1_M] \neq [h]$ since $\mu(N_h) > \mu(1_M)$. Therefore, $\psi = [h]$ is not a minimal projection.

Since ψ was an arbitrary non-zero projection, A_c has no minimal projections. $\qquad\square$

Combining the above results, we can also determine the minimal projections in the mixed subalgebra.

Proposition 5.8 *Let* $1 \leq j \leq \aleph_0$. *Then* $P_m(A_c \oplus A_d(j)) = \{(0, \delta_n) : n \in \underline{j}\}$.

Proof Suppose $(p, q) \in A_c \oplus A_d(j)$ is a projection. Then we can compute that $(p, q) = (p, q)^2 = (p^2, q^2)$, so $p^2 = p$ and $q^2 = q$. By the same reasoning, $p^* = p$ and $q^* = q$, whence $p \in A_c$ and $q \in A_d(j)$ are both projections. Since the converse is trivial, we conclude that the projections in $A_c \oplus A_d(j)$ are exactly formed by pairs of projections (p, q).

Now suppose (p, q) is a non-zero projection in $A_c \oplus A_d(j)$. Suppose $p \neq 0$. Then, since $p \in A_c$ and A_c has no minimal projections, there is a non-zero projection $p' \neq p$ in A_c such that $0 \leq p' \leq p$. Then $0 \leq (p', q) \leq (p, q)$, but we have $(p', q) \neq 0$ and $(p', q) \neq (p, q)$. Hence (p, q) is not a minimal projection.

Therefore, minimal projections in $A_c \oplus A_d(j)$ are necessarily of the form $(0, q)$, where q is a projection in $A_d(j)$. Since clearly $(p', q') \leq (p, q)$ if and only if $p' \leq p$ and $q \leq q'$, we see that $(0, q)$ is a minimal projection in $A_c \oplus A_d(j)$ if and only if q is a minimal projection in $A_d(j)$. Using Proposition 5.4 we therefore see that the minimal projections in $A_c \oplus A_d(j)$ are exactly given by $\{(0, \delta_n) : n \in \underline{j}\}$. $\qquad\square$

Hence we see that A_c is qualitatively different from $A_d(j)$ and from $A_c \oplus A_d(j)$ for some $1 \leq j \leq \aleph_0$, since the first does not contain any minimal projections, whereas the latter two do. Moreover, we can distinguish the discrete and the mixed subalgebras when considering the von Neumann algebra generated by the minimal projections. It is clear from Proposition 5.8 that the von Neumann algebra generated by the minimal projections in the mixed algebra is a subalgebra of $0 \oplus A_d(j)$ and is then certainly not equal to the whole mixed subalgebra itself. At the same time, we have the following statement about the discrete subalgebra. Note that $\langle X \rangle_{vN}$ denotes the von Neumann algebra generated by the set X, as discussed in Sect. B.4.

Proposition 5.9 *Let $1 \leq j \leq \aleph_0$. Then $\langle P_m(A_d(j)) \rangle_{vN} = A_d(j)$.*

Proof The minimal projections in $A_d(j)$ are exactly $\{\delta_n : n \in \underline{j}\}$, by Proposition 5.4. Now make a distinction between $j \in \mathbb{N}$ and $j = \aleph_0$. If $j \in \mathbb{N}$ and $f \in A_d(j)$, then

$$f = \sum_{n=1}^{j} f(n)\delta_n \in \langle \{\delta_n : n \in \underline{j}\} \rangle_{vN},$$

since a von Neumann algebra is closed under taking finite linear combinations. Hence

$$A_d(j) \subseteq \langle \{\delta_n : n \in \underline{j}\} \rangle_{vN},$$

if $j \in \mathbb{N}$. We now prove the same statement for $j = \aleph_0$. In this case $A_d(j) = \ell^\infty(\mathbb{N})$. So, take a $f \in \ell^\infty(\mathbb{N})$ and define $f_m = \sum_{n=1}^{m} f(n)\delta_n$ for all $m \in \mathbb{N}$.

Then certainly $f_m \in \langle \{\delta_n : n \in \underline{j}\} \rangle_{vN}$ for all $m \in \mathbb{N}$. Now let $\varphi \in \ell^2(\mathbb{N})$ and observe that

$$\|M_f(\varphi) - M_{f_m}(\varphi)\|^2 = \sum_{n=m+1}^{\infty} |f(n)\varphi(n)|^2 \leq \|f\|_\infty \sum_{n=m+1}^{\infty} |\varphi(n)|^2.$$

Since $\varphi \in \ell^2(\mathbb{N})$, it therefore follows that $\lim_{m \to \infty} \|M_f(\varphi) - M_{f_m}(\varphi)\| = 0$, i.e.

$$\lim_{m \to \infty} M_{f_m}(\varphi) = M_f(\varphi).$$

Since $\varphi \in \ell^2(\mathbb{N})$ was arbitrary, it follows that f is the strong limit of $\{f_m\}_{m=1}^{\infty}$, whence $f \in \langle \{\delta_n : n \in \mathbb{N}\} \rangle_{vN}$. Therefore,

$$A_d(j) \subseteq \langle \{\delta_n : n \in \underline{j}\} \rangle_{vN}$$

if $j = \aleph_0$ too.

Since $A_d(j)$ is a von Neumann algebra containing $\{\delta_n : n \in \underline{j}\}$, we have

$$\langle \{\delta_n : n \in \underline{j}\} \rangle_{vN} \subseteq A_d(j),$$

whence $A_d(j) = \langle \{\delta_n : n \in \underline{j}\} \rangle_{vN}$.   □

So, we can distinguish our three examples (the discrete, continuous and mixed sub-algebras) by considering minimal projections and the question whether they generate the whole algebra. Note that these two properties together divide up the collection of maximal abelian subalgebras in three classes:

- There are no minimal projections (like A_c),
- There are minimal projections that do not generate the whole algebra (as in the case of $A_c \oplus A_d(j)$),
- There are minimal projections that do generate the whole algebra (like $A_d(j)$).

In fact, this turns out to be the key to the classification of maximal abelian subalgebras.

5.3 Subalgebras Without Minimal Projections

We will first focus on the maximal abelian subalgebras that are like the continuous subalgebra, i.e. those that have no minimal projections. Our goal is to show that such subalgebras are unitarily equivalent to A_c. First of all, we need two definitions of special vectors.

Definition 5.10 Suppose H is a Hilbert space and $A \subseteq B(H)$ a C*-subalgebra. Then we say that $x \in H$ is a **separating vector** for A if $u \in A$ and $u(x) = 0$ implies that $u = 0$.

Definition 5.11 Suppose H is a Hilbert space and $A \subseteq B(H)$ a C*-subalgebra. Then we say that $x \in H$ is a **generating vector** for A if $\overline{Ax} = H$.

For maximal abelian subalgebras, it turns out that there is always a vector that is both generating and separating.

Proposition 5.12 *Suppose H is a separable Hilbert space and $A \subseteq B(H)$ is a maximal abelian von Neumann algebra. Then there is a unit vector $x \in H$ that is separating and generating for A.*

Proof We call a subset $C \subseteq H$ *orthogonal under A* if it has the property that $\{\overline{Ax}\}_{x \in C}$ is an orthogonal family (see Definition B.8). These subsets form a partially ordered set under inclusion and any chain $\{C_i\}_{i \in I}$ is bounded by $\bigcup_{i \in I} C_i$. Therefore, we can apply Zorn's lemma and obtain a maximal subset $E \subseteq H$ that is orthogonal under A.

Now note that $K := \bigoplus_{x \in E} \overline{Ax}$ is a closed subspace of H. Suppose $y \in K^\perp$. Then for $u, v \in A$ and $x \in E$, we have

$$\langle u(y), v(x) \rangle = \langle y, (u^*v)(x) \rangle = 0,$$

since $y \in K^\perp$ and $u^*v \in A$. Therefore, $\{u(y)\}$ is orthogonal to $\{v(x)\}$. Since $u, v \in A$ were arbitrary, $A(y)$ and $A(x)$ are orthogonal. By continuity of the inner product, therefore $\overline{A(y)}$ and $\overline{A(x)}$ are orthogonal. Since $x \in E$ was arbitrary, this means that $E \cup \{y\}$ is orthogonal under A.

However, by maximality of E, it follows that $y \in E$. Since $1 \in A$, $y \in \overline{Ay} \subseteq K$, so $y \in K \cap K^{\perp}$. Therefore, $y = 0$. So $K^{\perp} = \{0\}$, i.e. $K = H$.

Since H is separable, we know that E is (at most) countable. Furthermore, by maximality of E we know that $0 \in E$. Since removing 0 from E and normalizing the rest of E does not change the above properties, we can therefore find a subset $F = \{x_n \in H : n \in \mathbb{N}\} \subseteq H$ that consists of unit vectors, is orthogonal under A, and satisfies $\bigoplus_{n \in \mathbb{N}} \overline{Ax_n} = H$.

Now define $x := \sum_{n \in \mathbb{N}} 2^{-n} x_n$. Then, since $x_n \in \overline{Ax_n}$ for every $n \in \mathbb{N}$, we see that $\langle x_n, x_m \rangle = 0$ if $n \neq m$, so $\|x\|^2 = \sum_{n \in \mathbb{N}} 2^{-n} = 1$, i.e. x is a unit vector. We claim that x is both separating and generating for A.

For the first, suppose that $u \in A$ such that $u(x) = 0$. Then:

$$0 = \|u(x)\|^2 = \langle u(x), u(x) \rangle = \sum_{n,m \in \mathbb{N}} \frac{1}{2^{n+m}} \langle u(x_n), u(x_m) \rangle$$

$$= \sum_{n \in \mathbb{N}} \frac{1}{2^{2n}} \langle u(x_n), u(x_n) \rangle = \sum_{n \in \mathbb{N}} \frac{1}{2^{2n}} \|u(x_n)\|^2,$$

where we used the fact that F is orthogonal under A. Therefore, for each $n \in \mathbb{N}$, we have $\|u(x_n)\| = 0$, i.e. $u(x_n) = 0$. Now, for any $v \in A$, we obtain that $u(v(x_n)) = v(u(x_n)) = v(0) = 0$, since A is abelian, so $u(y) = 0$ for all $y \in A(x_n)$. So $u(y) = 0$ for all $y \in \overline{A(x_n)}$, so $u(y) = 0$ for every $y \in \bigoplus_{n \in \mathbb{N}} \overline{A(x_n)} = H$. Therefore, $u = 0$ and indeed x is a separating vector for A.

To see that x is a generating vector for A, denote $D := \overline{Ax}$. Since $D \subseteq H$ is closed, $H = D \oplus D^{\perp}$. Let π be the canonical projection from H onto D, i.e.

$$\pi : H \to H, (w, z) \mapsto (w, 0).$$

A is unital, so $x \in D$, whence $\pi(x) = x$ and $(1 - \pi)(x) = 0$.

We claim that $1 - \pi \in A$. To see this, note that for any $u, v \in A$,

$$u(v(x)) = (uv)(x) \in A(x),$$

so $u(Ax) \subseteq Ax$. By continuity of u, then also $u(D) \subseteq D$. Furthermore, if $y \in D^{\perp}$, then $y \in (Ax)^{\perp}$, so for any $u, v \in A$, $\langle u(y), v(x) \rangle = \langle y, (u^*v)(x) \rangle = 0$ and therefore $u(y) \in (Ax)^{\perp}$. Then by continuity of the inner product, $u(y) \in D^{\perp}$, too.

Hence every $a \in A$ splits in $(a_1, a_2) : D \oplus D^{\perp} \to D \oplus D^{\perp}$. Then for any $a \in A$,

$$a\pi = (a_1, a_2)(1, 0) = (a_1, 0) = (1, 0)(a_1, a_2) = \pi a,$$

i.e. $\pi \in A' = A$, since A is maximal abelian. Then also $1 - \pi \in A$, because A is an algebra.

So $1 - \pi \in A$ and $(1 - \pi)(x) = 0$, while x is a separating vector for A, so $1 - \pi = 0$, i.e. $\pi = 1$. Therefore,

$$\overline{Ax} = D = H,$$

and x is indeed a generating vector for A. $\square$

A von Neumann algebra has the special property that it is generated by its projections (viz. Proposition B.32). When it is also maximal abelian, there is an even stronger statement.

Lemma 5.13 *Suppose H is a separable Hilbert space and $A \subseteq B(H)$ is a maximal abelian von Neumann algebra. Then there is a countable set of projections in A that generates A as a von Neumann algebra.*

Proof By Proposition 5.12 there is a separating and generating vector $x \in H$ for A. Now let $D = \{px : p \in P(A)\}$. Since D is a subspace of the separable topological space H and is therefore also separable itself, D has a countable dense subspace

$$F = \{p_n x : n \in \mathbb{N}, \ p_n \in P(A)\}.$$

Now let $p \in P(A)$. Then $p(x) \in D$, so there is a sequence $\{n(i)\}_{i \in \mathbb{N}}$ such that

$$p(x) = \lim_{i \to \infty} p_{n(i)}(x).$$

Then for any $a \in A$,

$$pa(x) = ap(x) = a\left(\lim_{i \to \infty} p_{n(i)}(x)\right) = \lim_{i \to \infty} ap_{n(i)}(x) = \lim_{i \to \infty} p_{n(i)}a(x).$$

Now let $y \in H$ be arbitrary. Since x is a generating vector, $\overline{Ax} = H$, so there is a sequence $\{a_j\}_{j \in \mathbb{N}} \subseteq A$ such that $y = \lim_{j \to \infty} a_j(x)$. Then:

$$p(y) = p(\lim_{j \to \infty} a_j(x)) = \lim_{j \to \infty} p(a_j(x)) = \lim_{j \to \infty} \lim_{i \to \infty} p_{n(i)}(a_j(x))$$

$$= \lim_{i \to \infty} p_{n(i)}(\lim_{j \to \infty} a_j(x)) = \lim_{i \to \infty} p_{(n(i)}(y).$$

Therefore, p is the strong limit of $p_{n(i)}$. Since p was arbitrary,

$$P(A) \subseteq \langle\{p_n : n \in \mathbb{N}\}\rangle_{vN}.$$

Since $\langle P(A)\rangle_{vN} = A$ by Proposition B.32, we then have that

$$A \subseteq \langle\{p_n : n \in \mathbb{N}\}\rangle_{vN}.$$

However, A is a von Neumann algebra and $\{p_n : n \in \mathbb{N}\} \subseteq A$, so in fact we have

$$A = \langle\{p_n : n \in \mathbb{N}\}\rangle_{vN}. \qquad \square$$

Using Lemma 5.13, we can construct another special subset of the projections in the subalgebra. This one is no longer countable, but it has a lot more structure.

Lemma 5.14 *Suppose H is a separable Hilbert space and $A \subseteq B(H)$ is a maximal abelian von Neumann algebra. Then there is a maximal totally ordered family of projections in A that generates A as a von Neumann algebra.*

Proof By Lemma 5.13, we know that there is a countable set of projections $\{p_n\}_{n \in \mathbb{N}}$ in A that generates A as a von Neumann algebra. We claim that for every $n \in \mathbb{N}$ there is a finite, totally ordered set F_n of projections such that $F_n \subseteq F_{n+1}$ for all $n \in \mathbb{N}$ and the linear span of F_n contains p_n. We prove this by induction.

For our induction basis $n = 1$, take $F_1 = \{0, p_1, 1\}$.

Next, as our induction step, suppose that such an F_k has been constructed for all $k \leq n$. Since F_n is finite, totally ordered and contains F_1, $F_n = \{q_0, \ldots, q_r\}$ for some projections $0 = q_0 < q_1 < \cdots < q_r = 1$.

Now define $s_j = q_{j+1} - q_j$ for all $j \in \{1, \ldots, r\}$. Since $q_{j+1} > q_j$, s_j is again a projection in A, and satisfies $q_j s_j = 0$. Define:

$$F_{n+1} = F_n \cup \{q_j + s_j p_{n+1} : j \in \{0, \ldots, r-1\}\}.$$

First of all, note that for all $j \in \{0, \ldots, r-1\}$, $q_j + s_j p_{n+1}$ is a projection, since

$$(q_j + s_j p_{n+1})^* = q_j^* + p_{n+1}^* s_j^* = q_j + p_{n+1} s_j = q_j + s_j p_{n+1},$$

because A is abelian, and

$$\begin{aligned}
(q_j + s_j p_{n+1})^2 &= q_j^2 + q_j s_j p_{n+1} + s_j p_{n+1} q_j + s_j p_{n+1} s_j p_{n+1} \\
&= q_j + q_j s_j p_{n+1} + s_j^2 p_{n+1}^2 \\
&= q_j + s_j p_{n+1}.
\end{aligned}$$

So F_{n+1} consists of projections and is finite by construction.

Clearly, $s_j p_{n+1}$ is a projection for every $j \in \{0, \ldots, r-1\}$, whence $s_j p_{n+1} \geq 0$, so $q_j \leq q_j + s_j p_{n+1}$. Furthermore, note that $1 - p_{n+1}$ is a projection in A, too. Therefore, $s_j(1 - p_{n+1})$ is a projection in A, so is certainly positive. Hence

$$q_{j+1} - (q_j + s_j p_{n+1}) = s_j - s_j p_{n+1} = s_j(1 - p_n) \geq 0,$$

so $q_j + s_j p_{n+1} \leq q_{j+1}$. Therefore,

$$q_0 \leq q_0 + s_0 p_{n+1} \leq q_1 \leq q_1 + s_1 p_{n+1} \leq q_2 \leq \cdots \leq q_{r-1} \leq q_{r-1} + s_{r-1} p_{n+1} \leq q_r,$$

i.e. F_{n+1} is totally ordered.

By construction, $F_n \subseteq F_{n+1}$, so the only thing left to prove is that p_{n+1} is in the linear span of F_{n+1}. To see this, denote the linear span of F_{n+1} by V. Then for any $j \in \{0, \ldots, r\}$, $q_j \in V$ and $q_j + s_j p_{n+1} \in V$, so $s_j p_{n+1} = (q_j + s_j p_{n+1}) - q_j \in V$, since V is linear. Now observe that

$$\sum_{j=0}^{r-1} s_j = \sum_{j=0}^{r-1}(q_{j+1} - q_j) = \sum_{j=1}^{r} q_j - \sum_{j=0}^{r-1} q_j = q_r - q_0 = 1 - 0 = 1.$$

Therefore, $p_{n+1} = \sum_{j=0}^{r-1} s_j p_{n+1} \in V$. So, we have proven our induction step and have therefore proven our claim.

Now define $F_\infty = \bigcup_{n \in \mathbb{N}} F_n$. For any $q, q' \in F_\infty$ there are $l, m \in \mathbb{N}$ such that $q \in F_l$ and $q' \in F_m$, whence $q, q' \in F_{\max(l,m)}$, so either $q \leq q'$ or $q' \leq q$. Therefore, F_∞ is a totally ordered set of projections in A as well.

Now consider totally ordered sets G of projections in A that contain F_∞. The collection of all such G is endowed with a canonical partial order given by inclusion. Suppose

$$G_1 \subseteq G_2 \subseteq G_3 \subseteq \ldots$$

is a chain in this partial order. Then $\bigcup_{n \in \mathbb{N}} G_n$ again contains F_∞ and is totally ordered, by the same argument as the one used to show that F_∞ was totally ordered. Therefore, $\bigcup_{n \in \mathbb{N}} G_n$ is a member of the collection that we consider, i.e. every chain has an upper bound. Therefore, this collection has a maximal element F by Zorn's lemma.

For all $n \in \mathbb{N}$, p_n is in the linear span of F_n, so p_n is in the linear span of F_∞ and hence p_n is also in the linear span of F. Since $\langle \{p_n : n \in \mathbb{N}\} \rangle_{vN} = A$ by construction, $A \subseteq \langle F \rangle_{vN}$, but A is a von Neumann algebra and $F \subseteq A$, so $A = \langle F \rangle_{vN}$.

Therefore, F is a maximal totally ordered family of projections in A that generates A as a von Neumann algebra. $\qquad\square$

Using this maximal totally ordered family of projections and the properties of the projection lattice of a Hilbert space, we can prove the following rather technical but decisive result.

Proposition 5.15 *Suppose H is a separable Hilbert space and A is a maximal abelian von Neumann algebra without minimal projections. Furthermore, suppose that F is a maximal totally ordered set of projections in A and suppose that x is a generating and separating unit vector for A. Then the map $\psi : F \to [0, 1]$, given by $\psi(p) = \langle px, x \rangle$, is an isomorphism of partially ordered sets.*

Proof First of all, ψ is well-defined, since $0 \leq \langle px, x \rangle \leq 1$, by positivity of each projection $p \in F$ and the Cauchy-Schwarz inequality.

Now suppose that $p, q \in F$ such that $\psi(p) = \psi(q)$. Since F is totally ordered, we can assume that $p \leq q$. Then $q - p$ is also a projection in A, so

$$\|(q - p)(x)\|^2 = \langle (q - p)(x), (q - p)(x) \rangle = \langle (q - p)(x), x \rangle$$
$$= \langle q(x), x \rangle - \langle p(x), x \rangle = \psi(q) - \psi(p) = 0,$$

i.e. $(q - p)(x) = 0$. However, x is a generating vector for A, so $q - p = 0$. So $q = p$, i.e. ψ is injective. By the same computation, it is clear that for any $p \leq q$, we have $\psi(q) - \psi(p) = \|(q - p)(x)\|^2 \geq 0$, so $\psi(p) \leq \psi(q)$. Therefore, ψ is order preserving.

So, the only thing left to prove is that ψ is surjective. To see this, let $t \in [0, 1]$. Define:

$$F_0 := \{p \in F : \psi(p) < t\},$$

$$F_1 := \{p \in F : \psi(p) \geq t\}.$$

Clearly, F is the disjoint union of F_0 and F_1. Define $p_0 = \vee F_0$ and $p_1 = \wedge F_1$. By Proposition A.9, $p_0, p_1 \in F$.

Note that $p_0 \in \mathrm{Cl}_{\mathrm{str}}(F_0)$ by Proposition C.11, so for every $\varepsilon > 0$ there is a $p \in F_0$ such that $\psi(p_0) - \psi(p) = \|(p_0 - p)(x)\| < \varepsilon$. Therefore, $\psi(p_0) < \psi(p) + \varepsilon < t + \varepsilon$. Since $\varepsilon > 0$ is arbitrary, $\psi(p_0) \leq t$.

Likewise, $p_1 \in \mathrm{Cl}_{\mathrm{str}}(F_1)$, so for every $\varepsilon > 0$ there is a $q \in F_1$ such that

$$\psi(q) - \psi(p_1) = \|(q - p_1)(x)\| < \varepsilon,$$

i.e. $\psi(p_1) > \psi(q) - \varepsilon \geq t - \varepsilon$, whence $\psi(p_1) \geq t$.

So, we have the inequalities $\psi(p_0) \leq t \leq \psi(p_1)$. Since ψ is order preserving, we conclude that $p_0 \leq p_1$. Then $p_1 - p_0$ is a projection, so if $p_1 \neq p_0$, then there is a projection $q \in B(H)$ such that $0 \leq q \leq p_1 - p_0$, but neither $q = 0$ nor $q = p_1 - p_0$. Then also $p_0 \leq q + p_0 \leq p_1$, $p_0 \neq q + p_0$ and $q + p_0 \neq p_1$. Since $p_0 = \vee F_0$, then $q + p_0 \notin F_0$, and since $p_1 = \wedge F_1$, $q + p_0 \notin F_1$. Hence $q + p_0 \notin F$. However, for every $r \in F_0$, $r \leq p_0 \leq q + p_0$, and for every $s \in F_1$, $q + p_0 \leq p_1 \leq s$, so $F \cup \{q + p_0\}$ is totally ordered. This contradicts the maximality of F, so $p_1 = p_0$.

Then $\psi(p_0) \leq t \leq \psi(p_1) = \psi(p_0)$, i.e. $\psi(p_0) = t$. Since $t \in [0, 1]$ was arbitrary, ψ is surjective. Hence ψ is an isomorphism of ordered sets. □

Now, we are able to prove our main goal: whenever a maximal abelian subalgebra has no minimal projections, it is unitarily equivalent to the continuous subalgebra.

Theorem 5.16 *Suppose H is a separable Hilbert space and $A \subseteq B(H)$ is a maximal abelian von Neumann algebra that has no minimal projections. Then A is unitarily equivalent to A_c.*

Proof By Proposition 5.12 there is a separating and generating unit vector $x \in H$ for A. Furthermore, by Lemma 5.14, there is a maximal totally ordered family of projections F such that $\langle F \rangle_{vN} = A$. Combining these, by Proposition 5.15, the map $\varphi : F \to [0, 1]$, given by $\varphi(p) = \langle px, x \rangle$ is an isomorphism of ordered sets.

Now write $q_t = \varphi^{-1}(t) \in F$ for all $t \in [0, 1]$. Then $\langle q_t x, x \rangle = t$ for all $t \in [0, 1]$.

Furthermore, let $\chi_t : [0, 1] \to \mathbb{C}$ be the characteristic function of the interval $[0, t]$, where $t \in [0, 1]$. Then $[\chi_t] \in L^2(0, 1)$ for all $t \in [0, 1]$.

We now claim that there is a unique $u \in B(H, L^2(0, 1))$ such that $u(q_s x) = [\chi_s]$ for all $s \in [0, 1]$. To see this, first observe that $q_s q_t = q_{\min(s,t)}$ for all $s, t \in [0, 1]$ by construction. Therefore, for $s, t \in [0, 1]$,

$$\langle q_s x, q_t x \rangle = \langle q_t q_s x, x \rangle = \langle q_{\min(s,t)} x, x \rangle = \min(s, t),$$

and also

$$\langle [\chi_s], [\chi_t] \rangle = \int_{[0,1]} \overline{\chi_s(x)} \chi_t(x) \, \mathrm{d}x = \min(s, t).$$

Using this, we obtain:

$$\left\| \sum_{r=1}^{n} \mu_r q_{s_r} x \right\|^2 = \left\langle \sum_{r=1}^{n} \mu_r q_{s_r} x, \sum_{m=1}^{n} \mu_m q_{s_m} x \right\rangle = \sum_{r=1}^{n} \sum_{m=1}^{n} \overline{\mu_r} \mu_m \langle q_{s_r} x, q_{s_m} x \rangle$$

$$= \sum_{r=1}^{n} \sum_{m=1}^{n} \overline{\mu_r} \mu_m \langle [\chi_{s_r}], [\chi_{s_m}] \rangle = \left\langle \sum_{r=1}^{n} \mu_r [\chi_{s_r}], \sum_{m=1}^{n} \mu_m [\chi_{s_m}] \right\rangle$$

$$= \left\| \sum_{r=1}^{n} \mu_r [\chi_{s_r}] \right\|^2,$$

for any $\{\mu_r\}_{r=1}^{n} \subseteq \mathbb{C}$ and $\{s_r\}_{r=1}^{n} \subseteq [0, 1]$.

Now write S for the linear span of $\{q_s x : s \in [0, 1]\}$. By the above computation, if we have $\sum_r \mu_r q_{s_r} x = \sum_m \lambda_m q_{s_m} x \in S$, then

$$0 = \left\| \sum_r \mu_r q_{s_r} x - \sum_m \lambda_m q_{s_m} x \right\| = \left\| \sum_r \mu_r [\chi_{s_r}] - \sum_m \lambda_m [\chi_{s_m}] \right\|,$$

i.e. $\sum_r \mu_r [\chi_{s_r}] = \sum_m \lambda_m [\chi_{s_m}]$. Therefore, the map

$$v_1 : S \to L^2(0, 1), \quad \sum_r \mu_r q_{s_r} x \mapsto \sum_r \mu_r [\chi_{s_r}]$$

is well defined. By construction, v_1 is also linear, and by the above computations, we have $\|v_1(y)\| = \|y\|$ for any $y \in S$, so v_1 is certainly bounded. Lastly, by construction, $v_1(q_s x) = [\chi_s]$ for every $s \in [0, 1]$.

Since $\langle F \rangle_{vN} = A$, S is dense in Ax. Therefore, there is a unique bounded linear map $v_2 : Ax \to L^2(0, 1)$ that extends v_1. Then certainly $v_2(q_s x) = v_1(q_s x) = [\chi_s]$ for each $s \in [0, 1]$.

However, x is a generating vector for $A = A'$, so Ax is dense in H. So there is a unique $u \in B(H, L^2(0, 1))$ that extends v_2. Then also $u(q_s x) = v_2(q_s x) = [\chi_s]$ for all $s \in [0, 1]$, i.e. u satisfies our requirements.

To see that u is the unique element of $B(H, L^2(0, 1))$ that satisfies $u(q_s x) = [\chi_s]$ for each $s \in [0, 1]$, suppose that $u' \in B(H, L^2(0, 1))$ is such an element. Then by linearity, $u'|_S = u|_S = v_1$. Since S is dense in Ax, then $u'|_{Ax} = v_2$ and since Ax is dense in H, $u' = u$.

Hence indeed there is a unique $u \in B(H, L^2(0, 1))$ such that $u(q_s x) = [\chi_s]$ for each $s \in [0, 1]$. We claim that u is unitary. First observe that for any $s, t \in [0, 1]$, we have

$$\langle u(q_s x), u(q_t x) \rangle = \langle [\chi_s], [\chi_t] \rangle = \min(s, t) = \langle q_s x, q_t x \rangle,$$

so by linearity, $\langle uy, uz \rangle = \langle y, z \rangle$ for all $y, z \in H$. However, S is dense in H, so we have $\langle uy, uz \rangle = \langle y, z \rangle$ for all $y, z \in H$, i.e. u is indeed unitary.

Now observe that $[\chi_t] \in L^\infty(0, 1)$ too. Using this, we can compute, for any $s, t \in [0, 1]$:

$$(uq_s)(q_t x) = (u(q_s q_t))(x) = u(q_{\min(s,t)} x) = [\chi_{\min(s,t)}]$$
$$= [\chi_s \chi_t] = M_{[\chi_s]}([\chi_t]) = M_{[\chi_s]}(u(q_t x)) = (M_{[\chi_s]} u)(q_t x).$$

Therefore, $(uq_s)(y) = (M_{[\chi_s]} u)(y)$ for each $y \in S$ and $s \in [0, 1]$. Since S is dense in H, $uq_s = M_{[\chi_s]} u$ for all $s \in [0, 1]$.

Hence $uq_s u^{-1} = M_{[\chi_s]} \in A_c$ for all $s \in [0, 1]$, so $uFu^{-1} \subseteq A_c$.

Since $\langle F \rangle_{vN} = A$ and A_c is a von Neumann algebra, then also $uAu^{-1} \subseteq A_c$.

Then we have $A \subseteq u^{-1} A_c u$. Now A is maximal abelian, so $A = u^{-1} A_c u$, i.e. A and A_c are unitarily equivalent. $\qquad\square$

5.4 Subalgebras with Minimal Projections

Since we are now done with the case where the maximal abelian subalgebra has no minimal projections, we can move on to the case where it does. We first have the following two results.

Lemma 5.17 *Suppose H is a Hilbert space and $A \subseteq B(H)$ a von Neumann algebra. Furthermore, let $p \in P_m(A)$, then $pAp = \mathbb{C}p$.*

Proof Suppose $q \in pAp$ is a projection. Then $q = pap$ for some $a \in A$ and therefore $q \in A$ and $q(H) \subseteq p(H)$, so $q \in A$ and $q \leq p$. However, $p \in P_m(A)$, so $q = 0$ or $q = p$, so $q \in \mathbb{C}p$.

Now note that pAp is a von Neumann algebra by Lemma B.33, whence we have $\langle P(pAp)\rangle_{vN} = pAp$, by Proposition B.32. However, $P(pAp) \subseteq \mathbb{C}p$ by the above argument, so

$$pAp = \langle P(pAp)\rangle_{vN} \subseteq \mathbb{C}p.$$

For the reverse inclusion, let $\lambda \in \mathbb{C}$. Then observe that $1 \in A$, whence $\lambda 1 \in A$. Therefore, $\lambda p = \lambda p^2 = p(\lambda 1)p \in pAp$. So $\mathbb{C}p \subseteq pAp$. $\qquad\square$

Corollary 5.18 *Suppose H is a Hilbert space and $A \subseteq B(H)$ an abelian von Neumann algebra. Furthermore, suppose $p \in P_m(A)$ and $a \in A$. Then there is a $\lambda \in \mathbb{C}$ such that $ap = \lambda p$.*

Proof Observe that $ap = ap^2 = pap \in pAp = \mathbb{C}p$, by Lemma 5.17 and since A is abelian. Therefore, there is a $\lambda \in \mathbb{C}$ such that $ap = \lambda p$. $\qquad\square$

Now we need another technical result about subalgebras and projections.

Lemma 5.19 *Suppose H is a Hilbert space, $x \in H$, and $A \subseteq B(H)$ is a C^*-subalgebra. Furthermore, let q be the projection onto $\overline{Ax}$. Then $q \in A'$.*

Proof $\overline{Ax}$ is closed, so $H = \overline{Ax} \oplus \overline{Ax}^{\perp}$. We will show that any $a \in A$ decomposes over this splitting, i.e. that $a = (a_1, a_2)$, with $a_1 : \overline{Ax} \to \overline{Ax}$ and $a_2 : \overline{Ax}^{\perp} \to \overline{Ax}^{\perp}$.
   First let $a \in A$ and $y \in Ax$, say $y = bx$. Then $ay = (ab)x \in Ax$.
   Now let $z \in \overline{Ax}$, then $z = \lim_{i \in I} y_i$ for some $y_i \in Ax$ for every $i \in I$. Then

$$az = a(\lim_{i \in I} y_i) = \lim_{i \in I} ay_i \in \overline{Ax},$$

since $ay_i \in Ax$ for all $i \in I$. Hence $a(\overline{Ax}) \subseteq \overline{Ax}$ for all $a \in A$.
   Next, suppose $a \in A$, $z \in \overline{Ax}^{\perp}$ and $y \in \overline{Ax}$. Then:

$$\langle y, az\rangle = \langle a^*y, z\rangle = 0,$$

since $a^* \in A$ and $y \in \overline{Ax}$, so $a^*y \in \overline{Ax}$ by the above. Hence $az \in \overline{Ax}^{\perp}$.
   Therefore $a(\overline{Ax}^{\perp}) \subseteq \overline{Ax}^{\perp}$, so indeed, every $a \in A$ decomposes over the direct sum $H = \overline{Ax} \oplus \overline{Ax}^{\perp}$.
   Now note that $q = (1, 0) : \overline{Ax} \oplus \overline{Ax}^{\perp} \to \overline{Ax} \oplus \overline{Ax}^{\perp}$. Therefore, for any $a \in A$,

$$a \circ q = (a_1, a_2) \circ (1, 0) = (a_1, 0) = (1, 0) \circ (a_1, a_2) = q \circ a,$$

so $q \in A'$, as desired. $\qquad\square$

For a maximal abelian subalgebra A this has an important corollary, since then $A' = A$.

Corollary 5.20 *Suppose H is a Hilbert space, $x \in H$, and $A \subseteq B(H)$ is a maximal abelian subalgebra. Furthermore, let q be the projection onto $\overline{Ax}$. Then $q \in A$.*

Now, combining the above results, we can prove that the set of minimal projections in a maximal abelian subalgebra has an important structure.

Proposition 5.21 *Suppose H is a Hilbert space and $A \subseteq B(H)$ is a maximal abelian subalgebra. Then $P_m(A)$ is an orthogonal family of one-dimensional projections.*

Proof Suppose that $p, q \in P_m(A)$. Then certainly we also have $p, q \in A$, so by applying Corollary 5.18 twice, we see that there are $\lambda, \mu \in \mathbb{C}$ such that we have $\mu q = pq = qp = \lambda p$, since A is abelian. Again since A is abelian, pq is a projection too, whence $\lambda^2 = \lambda$ and $\mu^2 = \mu$. Therefore $\lambda, \mu \in \{0, 1\}$ and we also see that $\lambda = 0$ if and only if $\mu = 0$, i.e. $\lambda = \mu \in \{0, 1\}$.

Therefore, either $\mu = 0$ and then $pq = 0$ or $q = p$. Hence $P_m(A)$ is an orthogonal family of projections. To see that in fact all projections are one-dimensional, let $p \in P_m(A)$ and let $x \in p(H)$ be a non-zero vector. Then let q be the projection onto $\overline{Ax}$. Then $q \in A$ by Corollary 5.20. Furthermore, for $y \in Ax$, say $y = ax$ with $a \in A$,

$$py = pax = apx = ax = y,$$

so $Ax \subseteq p(H)$.

Therefore, $q(H) = \overline{Ax} \subseteq \overline{p(H)} = p(H)$, so $q \leq p$. However, $q \neq 0$, since $1 \in A$, whence $x = 1x \in Ax \subseteq q(H)$. Since $p \in P_m(A)$, it now follows that $p = q$, i.e. we have $p(H) = \overline{Ax}$.

Now note that by Corollary 5.18, for every $a \in A$ there is a $\lambda \in \mathbb{C}$ such that $ap = \lambda p$. Then $ax = apx = \lambda px = \lambda x$, so $Ax \subseteq \mathbb{C}x$, i.e. $p(H) = \overline{Ax}$ is at most one-dimensional. Since $x \in p(H)$ is non-zero, $p(H)$ is one-dimensional. $\quad\square$

Applying the above result to the case where the Hilbert space is separable, one can even say more.

Proposition 5.22 *Suppose H is a separable Hilbert space and $A \subseteq B(H)$ a maximal abelian von Neumann algebra. Then $P_m(A)$ is countable.*

Proof For every $p \in P_m(A)$, choose a unit vector $x_p \in p(H)$. Since $P_m(A)$ is an orthogonal family, $\{x_p : p \in P_m(A)\}$ is an orthonormal set in H and all x_p are different. Therefore, $\#P_m(A) = \#\{x_p : p \in P_m(A)\} \leq \dim(H)$. Since H is separable, $0 \leq \dim(H) \leq \aleph_0$, and therefore, $P_m(A)$ is countable. $\quad\square$

Now we come to one of our main points: every maximal abelian subalgebra that is generated by its minimal projections is unitarily equivalent to the discrete subalgebra.

Theorem 5.23 *Suppose H is a separable Hilbert space and $A \subseteq B(H)$ is a maximal abelian von Neumann algebra that is generated by its minimal projections. Furthermore, let j be the cardinality of $P_m(A)$. Then A is unitarily equivalent to $A_d(j)$.*

Proof By Proposition 5.22 we know that $1 \leq j \leq \aleph_0$, so it follows that there is a bijection $\varphi : \underline{j} \rightarrow P_m(A)$. Denote $\varphi(n) = p_n \in P_m(A)$ for all $n \in \underline{j}$. Now let

$$L := \left\{ \sum_{r=1}^{N} \mu_r p_{n_r} : \mu_r \in \mathbb{C}, \ n_r \in \underline{j} \right\},$$

i.e. L is the linear subspace of A spanned by $P_m(A)$. Then L is in fact an algebra, since for $c_1 = \sum_r \mu_r p_{n_r}, c_2 = \sum_s \lambda_s p_{n_s} \in L$, we have:

$$c_1 c_2 = \sum_{r,s} \mu_r \lambda_s p_{n_r} p_{n_s} = \sum_{r,s} \mu_r \lambda_s \delta_{n_r n_s} p_{n_r} \in L.$$

Furthermore, $(\sum_r \mu_r p_{n_r})^* = \sum_r \overline{\mu_r} p_{n_r} \in L$, so L is a *-algebra. Hence $\mathrm{Cl}_{\mathrm{str}}(L)$ is a von Neumann algebra. Clearly, $P_m(A) \subseteq \mathrm{Cl}_{\mathrm{str}}(L)$, so $\langle P_m(A) \rangle_{vN} \subseteq \mathrm{Cl}_{\mathrm{str}}(L)$. Furthermore, $L \subseteq \langle P_m(A) \rangle_{vN}$, so $\mathrm{Cl}_{\mathrm{str}}(L) \subseteq \langle P_m(A) \rangle_{vN}$. Hence, we obtain that $\mathrm{Cl}_{\mathrm{str}}(L) = \langle P_m(A) \rangle_{vN} = A$.

Now, for all $n \in \underline{j}$, choose a unit vector $e_n \in p_n(H)$ and let K be the closed linear subspace spanned by $\{e_n\}_{n \in j}$. Then $H = K \oplus K^\perp$. Suppose $x \in K^\perp$. Then for all $n \in \underline{j}$, $p_n(x) = \lambda_n e_n$ for some $\lambda_n \in \mathbb{C}$, since $p_n(H)$ is one-dimensional. Then

$$\lambda_n = \langle e_n, \lambda e_n \rangle = \langle e_n, p_n x \rangle = \langle p_n e_n, x \rangle = \langle e_n, x \rangle = 0,$$

for all $n \in \underline{j}$. Hence $p_n(x) = 0$ for all $n \in \underline{j}$. So $\psi(x) = 0$ for all $\psi \in L$, so $\psi(x) = 0$ for all $\psi \in \mathrm{Cl}_{\mathrm{str}}(L) = A$. Since $1 \in A$, therefore $x = 1x = 0$.

Hence $K^\perp = \{0\}$, i.e. $K = H$. Since for every $n, m \in \underline{j}$ we also have that

$$\langle e_n, e_m \rangle = \langle p_n e_n, p_m e_m \rangle = \langle p_m p_n e_n, e_m \rangle = \delta_{mn} \langle p_n e_n, e_m \rangle = \delta_{mn} \langle e_n, e_m \rangle = \delta_{mn},$$

we see that $\{e_n\}_{n \in \underline{j}}$ is in fact a basis for H.

Now define $u : \ell^2(\underline{j}) \rightarrow H$ by $u(f) = \sum_{n \in \underline{j}} f(n) e_n$. Then clearly, u is linear, and for $f \in \ell^2(\underline{j})$:

$$\|u(f)\|^2 = \left\langle \sum_{n \in \underline{j}} f(n) e_n, \sum_{m \in \underline{j}} f(m) e_m \right\rangle = \sum_{n \in \underline{j}} |f(n)|^2 = \|f\|^2.$$

Therefore, $u \in B(\ell^2(\underline{j}), H)$. Furthermore, for $f, g \in \ell^2(\underline{j})$,

$$\langle u(f), u(g) \rangle = \left\langle \sum_{n \in \underline{j}} f(n) e_n, \sum_{m \in \underline{j}} g(m) e_m \right\rangle = \sum_{n \in \underline{j}} \overline{f(n)} g(n) = \langle f, g \rangle,$$

so u is unitary. Now we claim that

$$A = \{\sum_{n \in \underline{j}} f(n)p_n : f \in \ell^\infty(\underline{j})\}.$$

To see this, first suppose $x \in H$. Then $x = \sum_{n \in \underline{j}} \lambda(n)e_n$, for some $\lambda \in \ell^2(\underline{j})$, since $\{e_n\}_{n \in \underline{j}}$ is a basis of H. Then

$$(\sum_{n \in \underline{j}} p_n)(x) = \sum_{m,n \in \underline{j}} p_n(\lambda(m)e_m) = \sum_{n \in \underline{j}} \lambda(n)e_n = x,$$

i.e. $\sum_{n \in \underline{j}} p_n = 1$.

Now, for every $n \in \underline{j}$ and $a \in A$, $ap_n = \lambda_a(n)p_n$ for some $\lambda_a(n) \in \mathbb{C}$, by Corollary 5.18. Therefore,

$$a = a \cdot 1 = \sum_{n \in \underline{j}} ap_n = \sum_{n \in \underline{j}} \lambda_a(n)p_n,$$

while

$$|\lambda_a(n)| = |\lambda_a(n)| \|p_n\| = \|\lambda_a(n)p_n\| = \|ap_n\| \le \|a\| \|p_n\| = \|a\|,$$

whence

$$\sup_{n \in \underline{j}} |\lambda_a(n)| \le \|a\|,$$

i.e. $\lambda_a \in \ell^\infty(\underline{j})$. Therefore, $A \subseteq \{\sum_{n \in \underline{j}} f(n)p_n : f \in \ell^\infty(\underline{j})\}$.

Now let $f \in \ell^\infty(\underline{j})$ and $x \in H$. Then $x = \sum_{n \in \underline{j}} \lambda(n)e_n$, with $\lambda \in \ell^2(\underline{j})$, since $\{e_n\}_{n \in \underline{j}}$ is a basis for H. Then for all $m \in \underline{j}$, $\sum_{n \le m} f(n)p_n(x) = \sum_{n \le m} f(n)\lambda(n)e_n$, so

$$\|\sum_{n \le m} f(n)p_n(x)\|^2 = \sum_{n \le m} |f(n)|^2 |\lambda(n)|^2 \le \|f\|_\infty^2 \cdot \sum_{n \le m} |\lambda(n)|^2.$$

Hence $\sum_{n \in \underline{j}} f(n)p_n(x)$ is well defined and

$$\|\sum_{n \in \underline{j}} f(n)p_n(x)\| \le \|f\|_\infty \|\lambda\| = \|f\|_\infty \|x\|,$$

so $a := \sum_{n \in \underline{j}} f(n)p_n \in B(H)$. We now claim that $a \in A$. To see this, define for all $m \in \underline{j}$, $b_m = \sum_{n \le m} f(n)p_n \in A$. Further, let again be $x \in H$, where we write $x = \sum_{n \in \underline{j}} \lambda(n)e_n$, where $\lambda \in \ell^2(\underline{j})$. Then:

$$\|b_m(x) - a(x)\|^2 = \|\sum_{n>m} f(n)p_n(x)\|^2$$

$$= \langle \sum_{n>m} f(n)\lambda(n)e_n, \sum_{k>m} f(k)\lambda(k)e_k \rangle$$

$$= \sum_{n>m} |f(n)|^2 |\lambda(n)|^2$$

$$\leq \|f\|_\infty^2 \sum_{n>m} |\lambda(n)|^2.$$

Therefore, $\{b_m(x)\}_{m\in\underline{j}}$ converges to $a(x)$. Hence, $\{b_m\}_{m\in\underline{j}}$ converges to a in the strong topology. However, A is strongly closed, so $a \in A$. Therefore

$$A = \{\sum_{n\in\underline{j}} f(n)p_n : f \in \ell^\infty(\underline{j})\}.$$

Now suppose $f \in \ell^\infty(\underline{j})$ and $g \in \ell^2(\underline{j})$. Then:

$$(uM_f)(g) = u(M_f(g)) = \sum_{n\in\underline{j}} (M_f(g))(n)e_n$$

$$= \sum_{n\in\underline{j}} f(n)g(n)e_n = \sum_{n,m\in\underline{j}} f(n)g(m)p_n(e_m)$$

$$= (\sum_{n\in\underline{j}} f(n)p_n)(\sum_{m\in\underline{j}} g(m)e_m) = (\sum_{n\in\underline{j}} f(n)p_n)(u(g))$$

$$= ((\sum_{n\in\underline{j}} f(n)p_n)u)(g),$$

whence $uM_f = (\sum_{n\in\underline{j}} f(n)p_n)u$, i.e. $uM_f u^{-1} = \sum_{n\in\underline{j}} f(n)p_n \in A$.

Therefore, $uA_d(\underline{j})u^{-1} \subseteq A$, so $A_d(\underline{j}) \subseteq u^{-1}Au$. However, $A_d(\underline{j})$ is maximal abelian and $u^{-1}Au$ is abelian, so $A_d(\underline{j}) = u^{-1}Au$. Therefore, A is unitarily equivalent to $A_d(\underline{j})$. $\qquad\square$

Finally, there is the case that the maximal abelian subalgebra does have minimal projections, but is not generated by them.

Theorem 5.24 *Let H be a separable Hilbert space and $A \subseteq B(H)$ a maximal abelian von Neumann algebra. Furthermore, let $1 \leq j \leq \aleph_0$ and suppose that $\#P_m(A) = j$ and $\langle P_m(A)\rangle_{vN} \neq A$. Then A is unitarily equivalent to $A_d(j) \oplus A_c$.*

Proof By assumption, there is a bijection $\varphi : \underline{j} \to P_m(A)$. Now write $p_n = \varphi(n)$ and define $p = \sum_{n\in\underline{j}} p_n$. Since $P_m(A)$ is an orthogonal family, $p \in B(H)$ is again a projection. Since A is strongly closed and p is the strong limit of the net $\{\sum_{n\leq m} p_n\}_{m\in\underline{j}}$, $p \in A$. Since p is a projection, $K := p(H)$ is a closed linear subspace of H and therefore $H = K \oplus K^\perp$.

Now we claim that A decomposes over $K \oplus K^\perp$. To see this, let $a \in A$ and observe that $a = ap + a(1 - p)$. First let $x \in K$. Then

$$ax = apx + a(1 - p)x = p(ax) + 0 = p(ax) \in K,$$

where we used the assumption that A is abelian. Next, for $x \in K^\perp = (1 - p)(H)$,

$$ax = apx + a(1 - p)x = 0 + (1 - p)ax = (1 - p)(ax) \in K^\perp.$$

Therefore a decomposes over $K \oplus K^\perp$ and therefore indeed A decomposes over $K \oplus K^\perp$.

So, for every $a \in A$, there is a unique $a_1 \in B(K)$ and a unique $a_2 \in B(K^\perp)$ such that $a = (a_1, a_2)$. Now define:

$$A_1 := \{a|_K : a \in A, a_2 = 0\} \subseteq B(K),$$

and

$$A_2 := \{a|_{K^\perp} : a \in A, a_1 = 0\} \subseteq B(K^\perp).$$

Now we claim that $A = A_1 \oplus A_2$. To see this, first let $a \in A$. Then $a = (a_1, a_2)$ with $a_1 \in B(K)$ and $a_2 \in B(K^\perp)$. Then clearly $a_1 = (a_1, 0)|_K$ and we also have that $(a_1, 0) = (a_1, a_2)(1, 0) = ap \in A$, so $a_1 \in A_1$. Likewise, it follows that $a_2 \in A_2$, so $a = (a_1, a_2) \in A_1 \oplus A_2$. Hence $A \subseteq A_1 \oplus A_2$.

For the converse, suppose $b_1 \in A_1$ and $b_2 \in A_2$. Then $(b_1, 0) \in A$ and $(0, b_2) \in A$. Therefore, $(b_1, b_2) = (b_1, 0) + (0, b_2) \in A$. Therefore, $A_1 \oplus A_2 \subseteq A$ and hence we have $A = A_1 \oplus A_2$.

Now let $a, b \in A_1$. Then $(a, 0), (b, 0) \in A$, whence

$$(ab, 0) = (a, 0)(b, 0) = (b, 0)(a, 0) = (ba, 0),$$

since A is abelian. Therefore, $ab = ba$, i.e. A_1 is abelian. Likewise, A_2 is abelian.

Now suppose $A_1 \subseteq C \subseteq B(K)$ and C is an abelian subalgebra. Then

$$A = A_1 \oplus A_2 \subseteq C \oplus A_2 \subseteq B(H),$$

and A is maximal abelian, so $A = A_1 \oplus A_2 = C \oplus A_2$. Therefore, $A_1 = C$, and the subalgebra $A_1 \subseteq B(K)$ is maximal abelian. With a similar argument, $A_2 \subseteq B(K^\perp)$ is maximal abelian.

Now we claim that $P_m(A_1) = \{p_n|_K : n \in \underline{j}\}$. To see this, let $q \in P_m(A_1)$. Then $q \in A_1$, so $(q, 0) \in A$ and $(q, 0)$ is a projection. Now suppose $0 \leq s \leq (q, 0)$ for some projection $s \in A$. Then $s = (s_1, s_2)$ for some projections $s_1 \in B(K)$ and $s_2 \in B(K^\perp)$. Then $s_1 = (s_1, 0)|_K = (sp)|_K$ and $sp \in A$, so $s_1 \in A_1$. We then have $0 \leq s_1 \leq q$ and $0 \leq s_2 \leq 0$, so $s_2 = 0$ and $s_1 = 0$ or $s_1 = q$, since $q \in P_m(A_1)$. Therefore, either $s = 0$ or $s = (q, 0)$, whence $(q, 0) \in P_m(A)$. So, we know that there is a $n \in \underline{j}$ such that $(q, 0) = p_n = (p_n|_K, 0)$, i.e. $q = p_n|_K$.

For the converse, suppose that $n \in \underline{j}$ and that q is a projection in A_1, such that we have $0 \leq q \leq p_n|_K$. Then $(0,0) \leq (q,0) \leq (p_n|_K, 0) = p_n$, so either $(q,0) = 0$ or $(q,0) = (p_n|_K, 0)$, i.e. $q = 0$ or $q = p_n|_K$. Therefore, $p_n|_K \in P_m(A_1)$. So indeed, $P_m(A_1) = \{p_n|_K : n \in \underline{j}\}$.

The next thing we want to prove is $\langle P_m(A_1) \rangle_{vN} = A_1$. Clearly, $\langle P_m(A_1) \rangle_{vN} \subseteq A_1$, since A_1 is a von Neumann algebra. For the converse, suppose $a_1 \in A_1$. Then we have $a = (a_1, 0) \in A$, whence $ap = (a_1, 0)(1, 0) = (a_1, 0) = a$, so

$$(a_1, 0) = a = ap = a \sum_{n \in \underline{j}} p_n$$

$$= \sum_{n \in \underline{j}} ap_n = \sum_{n \in \underline{j}} \lambda_a(n) p_n$$

$$= \sum_{n \in \underline{j}} \lambda_a(n)(p_n|_K, 0) = \left(\sum_{n \in \underline{j}} \lambda_a(n) p_n|_K, 0\right).$$

Here we used Corollary 5.18 to find the $\lambda_a(n) \in \mathbb{C}$. Therefore,

$$a_1 = \sum_{n \in \underline{j}} \lambda_a(n) p_n \in \langle \{p_n|_K : n \in \underline{j}\} \rangle_{vN} = \langle P_m(A_1) \rangle_{vN}.$$

Hence, indeed, $\langle P_m(A_1) \rangle_{vN} = A_1$. So $A_1 \subseteq B(K)$ is a maximal abelian von Neumann algebra that is generated by its j minimal projections. Therefore, by Theorem 5.23, there is unitary $u_1 \in B(K, \ell^2(\underline{j}))$ such that

$$u_1 A_1 u_1^{-1} = A_d(\underline{j}).$$

Next, we claim that A_2 has no minimal projections. To see this, suppose that $q \in A_2$ is a non-zero projection. Then $(0, q) \in A$ is a projection, and

$$(0, q) \notin P_m(A) = \{p_n = (p_n|_K, 0) : n \in \underline{j}\}.$$

Therefore, there is a projection $s \in A$ such that $0 \leq s \leq (0, q)$ and $s \neq 0, s \neq (0, q)$. Then $s = (s_1, s_2) \in A_1 \oplus A_2$ for some projections $s_1 \in A_1$ and $s_2 \in A_2$. Then $0 \leq (s_1, s_2) \leq (0, q)$, so $s_1 = 0$ and $0 \leq s_2 \leq q$ with $s_2 \neq 0$ and $s_2 \neq q$. Therefore, $q \notin P_m(A_2)$. Since $q \in A$ was an arbitrary projection, $P_m(A_2) = \emptyset$.

Therefore, A_2 is a maximal abelian von Neumann algebra without minimal projections, so by Theorem 5.16 there is a unitary $u_2 \in B(K^\perp, L^2(0, 1))$ such that

$$u_2 A_2 u_2^{-1} = A_c.$$

Now, $(u_1, u_2) \in B(H, \ell^2(\underline{j}) \oplus L^2(0, 1))$ is a unitary such that

$$(u_1, u_2)A(u_1, u_2)^{-1} = (u_1, u_2)(A_1 \oplus A_2)(u_1^{-1}, u_2^{-1})$$
$$= u_1 A_1 u_1^{-1} \oplus u_2 A_2 u_2^{-1}$$
$$= A_d(j) \oplus A_c$$

i.e. A is unitary equivalent to $A_d(j) \oplus A_c$, as desired.   □

5.5 Classification

For a separable Hilbert space H and a maximal abelian von Neumann algebra, we have showed in the previous sections that $P_m(A)$ determines A up to unitary equivalence. More explicitly, we have the following result.

Corollary 5.25 *Suppose H is a separable Hilbert space and $A \subseteq B(H)$ is a maximal abelian von Neumann algebra. Then A is unitarily equivalent to exactly one of the following:*

1. $A_c \subseteq B(L^2(0, 1))$
2. $A_d(j) \subseteq B(\ell^2(\underline{j}))$ *for some* $1 \le j \le \aleph_0$
3. $A_d(j) \oplus A_c \subseteq B(\ell^2(\underline{j}) \oplus L^2(0, 1))$ *for some* $1 \le j \le \aleph_0$.

This classification has the following very important corollary for our main goal.

Corollary 5.26 *Suppose H is a separable Hilbert space and $A \subseteq B(H)$ a unital abelian subalgebra that has the Kadison-Singer property. Then A is unitarily equivalent to either $A_d(j)$ for some $1 \le j \le \aleph_0$, A_c or $A_d(j) \oplus A_c$ for some $1 \le j \le \aleph_0$.*

In the rest of this text, we will determine whether the discrete, continuous and mixed subalgebra have the Kadison-Singer property. So far, we only know that $A_d(j)$ has the Kadison-Singer property if $j \in \mathbb{N}$.

Chapter 6
Stone-Čech Compactification

We will first focus on the question whether the continuous subalgebra has the Kadison-Singer property. We will answer this question (negatively) in Chap. 7, but to do this, we first need to discuss the topological notion of a *Stone-Čech compactification*. This is a topological space that can be seen as the biggest compactification of a given topological space. Its universal property is useful in a wide variety of contexts and is also precisely what we will use in Chap. 7.

Not every topological space admits a Stone-Čech compactification, but so-called *Tychonoff spaces* do. In this chapter, we construct the Stone-Čech compactification for such spaces using ultrafilters on zero-sets.

6.1 Stone-Čech Compactification

Definition 6.1 Suppose X is a topological space. The **Stone-Čech compactification** of X is a pair $(\beta X, S)$, where βX is a compact Hausdorff space βX, and S is a continuous map $S : X \to \beta X$ having the following universal property: for any compact Hausdorff space K and continuous function $f : X \to K$, there is a unique continuous $\beta f : \beta X \to K$ such that the following diagram commutes:

The Stone-Čech compactification is unique up to homeomorphism. Therefore, we can speak of 'the' (rather than 'a') Stone-Čech compactification of a topological space. However, not every topological space admits a Stone-Čech compactification. However, so-called *Tychonoff spaces* do, as we show in what follows.

M. Stevens, *The Kadison-Singer Property*,
SpringerBriefs in Mathematical Physics 14, DOI 10.1007/978-3-319-47702-2_6

6.2   Ultrafilters

There a multiple constructions of the Stone-Čech compactification, but we will use the construction based on *ultrafilters*. We will define these for *meet-semilattices*, which can be seen as the most general setting for doing this.

Definition 6.2 A **meet-semilattice** is a partial order $(\Sigma, \leq)$ with the property that any two elements $x, y \in \Sigma$ have a **meet** z, i.e. there exists an element $z \in \Sigma$ such that $z \leq x$, $z \leq y$ and for all $w \in \Sigma$ such that both $w \leq x$ and $w \leq y$, we have $w \leq z$. We denote the meet of x and y by $x \wedge y$.

We will often denote a meet-semilattice by Σ instead of $(\Sigma, \leq)$ and imply that the order is given by the symbol $\leq$. For some purposes, meet-semilattices need to have some more structure.

Definition 6.3 A **lattice** is a meet-semilattice Σ with the property that any two elements $a, b \in \Sigma$ have a **join**, i.e. an element $c \in \Sigma$ with the property that $a \leq c$, $b \leq c$ and for any $d \in \Sigma$ such that both $a \leq d$ and $b \leq d$, we have $c \leq d$.

We can now define filters for meet-semilattices.

Definition 6.4 Suppose Σ is a meet-semilattice. A family $F \subseteq \Sigma$ is called a **filter** (for Σ) if it satisfies the following axioms:

1. $F \neq \emptyset$,
2. $F \neq \Sigma$,
3. if $a, b \in \Sigma$, then $a \wedge b \in F$ and
4. if $a \in F$ and $a \leq b$, then $b \in F$.

An ultrafilter is now just a special kind of filter.

Definition 6.5 Suppose Σ is a meet-semilattice and $F \subseteq \Sigma$ is a filter. Then F is called an **ultrafilter** (for Σ) if the only filter $G \subseteq \Sigma$ that satisfies $F \subseteq G$ is F itself. We denote all ultrafilters on Σ by Ultra(Σ).

The property of being maximal does characterize an ultrafilter, but it is not always the easiest to work with. Therefore, we introduce another description of ultrafilters in the special case that the meet-semilattice has a minimal element.

Lemma 6.6 *Suppose Σ is a meet-semilattice with a minimal element 0 and $F \subseteq \Sigma$ is a filter. Then F is an ultrafilter if and only if F has the following property: if $a \in \Sigma$ and $a \wedge b \neq 0$ for all $b \in F$, then $a \in F$.*

Proof First suppose that F is an ultrafilter. Furthermore, suppose that $a \in \Sigma$ such that $a \wedge b \neq 0$ for all $b \in F$. Then define

$$F' = F \cup \{c \in \Sigma \mid \exists b \in F : a \wedge b \leq c\}.$$

We leave the straightforward computation that F' is a filter to the reader. By construction, $F \subseteq F'$ and F is an ultrafilter, so $F' = F$. Now, take any $b \in F$. Then $a \wedge b \leq a$, so $a \in F' = F$.

For the converse, suppose that $a \in \Sigma$ and $a \wedge b \neq 0$ for all $b \in F$ imply that $a \in F$. Then suppose $G \subseteq \Sigma$ is a filter such that $F \subseteq G$. Then let $a \in G$. Then for any $b \in F$, $a, b \in G$, so $a \wedge b \in G$, so $a \wedge b \neq 0$. Therefore, $a \in F$, by our assumption. Hence $G \subseteq F$, i.e. $G = F$. Therefore, F is an ultrafilter. $\square$

In the case that a meet-semilattice has a minimal element, we can also prove the following.

Lemma 6.7 *Suppose Σ is a meet-semilattice with a minimal element. Then, for any filter F on Σ, there is an ultrafilter G on Σ such that $F \subseteq G$.*

Proof We prove this using Zorn's lemma, i.e. let F be a filter on Σ and consider the set $\mathscr{F}$ consisting of all filters G on Σ such that $F \subseteq G$. Then let $\{H_i\}_{i \in I}$ be a chain in $\mathscr{F}$. Then define $H = \bigcup_{i \in I} H_i$. Clearly, $F \subseteq H$. Furthermore, it is easy to see that H is a filter, since $\{H_i\}_{i \in I}$ is totally ordered. Hence $H \in \mathscr{F}$ and H is an upper bound of $\{H_i\}_{i \in I}$.

Therefore, $\mathscr{F}$ is a partially ordered set with the property that every chain has an upper bound. Hence, by Zorn's lemma, $\mathscr{F}$ has a maximal element G. To see that G is in fact an ultrafilter, suppose that G' is also a filter on Σ such that $G \subseteq G'$. Then $F \subseteq G'$, so $G' \in \mathscr{F}$. Since $G \subseteq G'$ and G is a maximal element of $\mathscr{F}$, then $G' = G$, i.e. G is an ultrafilter such that $F \subseteq G$. $\square$

For lattices, we can consider another class of filters.

Definition 6.8 Suppose Σ is a lattice and let F be a filter on Σ. Then F is called **prime** if $a \vee b \in F$ implies $a \in F$ or $b \in F$.

Prime filters become particularly interesting when the lattice is distributive.

Definition 6.9 Suppose Σ is a lattice. Σ is called **distributive** if for any $a, b, c \in \Sigma$, we have $a \vee (b \wedge c) = (a \vee b) \wedge (a \vee c)$ and $a \wedge (b \vee c) = (a \wedge b) \vee (a \wedge c)$.

For distributive lattice with a minimal elements, the prime filters are exactly the ultrafilters.

Lemma 6.10 *Suppose Σ is a distributive lattice with a minimal element 0. Then every ultrafilter on Σ is prime.*

Proof Suppose F is a an ultrafilter on Σ. Furthermore, suppose $c_1, c_2 \in \Sigma$ such that $c_1 \vee c_2 \in F$ and $c_1, c_2 \notin F$. Then, by Lemma 6.6, there are $b_1, b_2 \in F$, such that $c_1 \wedge b_1 = 0$ and $c_2 \wedge b_2 = 0$. Then

$$(c_1 \vee c_2) \wedge (b_1 \wedge b_2) = (c_1 \wedge b_1 \wedge b_2) \vee (c_2 \wedge b_1 \wedge b_2) \leq (c_1 \wedge b_1) \vee (c_2 \wedge b_2) = 0 \vee 0 = 0.$$

Therefore, $(c_1 \vee c_2) \wedge (b_1 \wedge b_2) = 0$, but both $c_1 \vee c_2 \in F$ and $b_1 \wedge b_2 \in F$, so then $0 \in F$. This is in contradiction with F being a filter, so F is prime. $\square$

6.3 Zero-Sets

We now turn our attention to one specific example of a meet-semilattice, namely the set of *zero-sets* of a topological space.

Definition 6.11 Suppose X is a topological space and let $A \subseteq X$. A is called a **zero-set** if there is a continuous function $f : X \rightarrow [0, 1]$ such that $A = f^{-1}(\{0\})$. The collection of all zero-sets in X is denoted by $Z(X)$.

Note that every zero-set is closed, since $\{0\} \subseteq [0, 1]$ is closed. Furthermore, for discrete spaces, we have $Z(X) = \mathscr{P}(X)$. The collection of zero-sets of a topological space has the following property:

Lemma 6.12 *Suppose X is a topological space and let $A, B \in Z(X)$ be zero-sets such that $A \cap B = \emptyset$. Then there are $C, D \in Z(X)$ such that*

$$A \subseteq X \setminus C \subseteq D \subseteq X \setminus B.$$

Proof Since $A, B \in Z(X)$, there are continuous functions $f, g : X \rightarrow [0, 1]$ such that $A = f^{-1}(\{0\})$ and $B = g^{-1}(\{0\})$. Now define $h : X \rightarrow [0, 1]$ by $h = \frac{f}{f+g}$. Then h is well defined (since $A \cap B = \emptyset$) and continuous. Note that we have $h^{-1}(\{0\}) = f^{-1}(\{0\}) = A$ and $h^{-1}(\{1\}) = g^{-1}(\{0\}) = B$. Now, let

$$C := \{x \in X \mid h(x) \geq \tfrac{1}{2}\},$$

$$D := \{x \in X \mid h(x) \leq \tfrac{1}{2}\}.$$

Then C is the zero-set of the continuous function $\max\{\tfrac{1}{2} - h, 0\}$, so $C \in Z(X)$, and likewise, $D \in Z(X)$. Furthermore, we clearly have that $A \subseteq X \setminus C$, $X \setminus C \subseteq D$ and $D \subseteq X \setminus B$. □

We now show that for a topological space X, $Z(X)$ is in fact a lattice.

Lemma 6.13 *Suppose X is a topological space. Then $Z(X)$ is a sublattice of $\mathscr{P}(X)$ that contains $\emptyset$ and X.*

Proof It is clear that $\emptyset$ and X are contained in $Z(X)$. To see that $Z(X)$ is in fact a lattice, suppose $A, B \in Z(X)$. Then there are continuous functions $f, g : X \rightarrow [0, 1]$ such that $f^{-1}(\{0\}) = A$ and $g^{-1}(\{0\}) = B$. Then $h := \frac{f+g}{2}$ gives $A \cap B$ as a zero-set and $k = f \cdot g$ gives $A \cup B$ as a zero-set. □

Lemma 6.13 guarantees that for any topological space X, we can consider ultrafilters on $Z(X)$. In fact, $\mathrm{Ultra}(Z(X))$ serves as the underlying set for the Stone-Čech compactification of X, whenever X is a Tychonoff space.

6.4 Ultra-Topology

For a topological space X, we want to endow $\text{Ultra}(Z(X))$ with a topology.

Definition 6.14 Suppose X is a topological space and $A \in Z(X)$. Then we define

$$U(A) = \{F \in \text{Ultra}(Z(X)) \mid A \notin F\},$$

and

$$W(A) = \text{Ultra}(Z(X)) \setminus U(A) = \{F \in \text{Ultra}(Z(X)) \mid A \in F\}.$$

We use the sets $U(A)$ to define a topology on $\text{Ultra}(Z(X))$. Namely, observe that for any $F \in \text{Ultra}(Z(X))$, we have that $F \neq Z(X)$, so there is an $A \in Z(X)$ such that $A \notin F$, i.e. $F \in U(A)$. Hence

$$\bigcup_{A \in Z(X)} U(A) = \text{Ultra}(Z(X)),$$

so $\{U(A) \mid A \in Z(X)\}$ is a subbase for a topology on $\text{Ultra}(Z(X))$. We will call this topology the **ultra-topology**.

From now on, we will consider $\text{Ultra}(Z(X))$ as a topological space, endowed with the ultra-topology. In order to better understand this topology, observe that the following identities hold.

For a topological space X and a subset $\{A_i\}_{i=1}^n \subseteq Z(X)$, we have

$$U\left(\bigcap_{i=1}^n A_i\right) = \bigcup_{i=1}^n U(A_i)$$

and

$$U\left(\bigcup_{i=1}^n A_i\right) = \bigcap_{i=1}^n U(A_i).$$

The first identity can be obtained by a direct computation using the defining relation $W(A) = \text{Ultra}(Z(X)) \setminus U(A)$, whereas the second is a corollary of the fact that every ultrafilter on $Z(X)$ is prime. These identities combined with the definition of the ultra-topology has the following immediate corollary.

Corollary 6.15 *Suppose X is a topological space. Then $\{U(A) \mid A \in Z(X)\}$ is a base for the ultra-topology on $\text{Ultra}(Z(X))$.*

As we mentioned in Sect. 6.3, $\text{Ultra}(Z(X))$ is the underlying set of the Stone-Čech compactification for a Tychonoff space X. Since we have now endowed $\text{Ultra}(Z(X))$ with the ultra-topology, we need to show that in fact this makes $\text{Ultra}(Z(X))$ a compact Hausdorff space. First of all, we prove that it is Hausdorff.

Proposition 6.16 *Suppose X is a topological space. Then* $\mathrm{Ultra}(Z(X))$ *is Hausdorff.*

Proof Suppose $F \neq G \in \mathrm{Ultra}(Z(X))$. Then, since F is maximal, there is an $A \in F$ such that $A \notin G$. Then, by Lemma 6.6, there is a $B \in G$ such that $A \cap B = \emptyset$.

Now, using Lemma 6.12, there are $C, D \in Z(X)$ such that

$$A \subseteq X \setminus C \subseteq D \subseteq X \setminus B.$$

Since $A \in F$, $C \in F$ would imply $\emptyset = A \cap C \in F$, so $C \notin F$, i.e. $F \in U(C)$. Likewise, since $B \in G$, $G \in U(D)$. Furthermore, $U(C), U(D) \subseteq \mathrm{Ultra}(Z(X))$ are open and

$$U(C) \cap U(D) = U(C \cup D) = U(X) = \emptyset.$$

Therefore, $\mathrm{Ultra}(Z(X))$ is Hausdorff. $\square$

Next, we prove that $\mathrm{Ultra}(Z(X))$ is compact. We do this in three steps.

Lemma 6.17 *Suppose X is a topological space and let $A \in Z(X)$. Then $W(A) = \emptyset$ if and only if $A = \emptyset$.*

Proof Suppose $A = \emptyset$. Then for every filter F on $Z(X)$, $\emptyset \notin F$, so for every ultrafilter G on $Z(X)$, $G \notin W(\emptyset)$, i.e. $W(\emptyset) = \emptyset$.

Next, suppose $A \neq \emptyset$. Then, clearly,

$$F_A := \{B \in Z(X) \mid A \subseteq B\}$$

is a filter on $Z(X)$ containing A. Then, by Lemma 6.7, there is an ultrafilter G_A such that $A \in F_A \subseteq G_A$. Hence $G_A \in W(A)$, i.e. $W(A) \neq \emptyset$.

So, indeed, $W(A) = \emptyset$ if and only if $A = \emptyset$. $\square$

Lemma 6.18 *Suppose X is a topological space and let $\{A_i\}_{i \in I} \subseteq Z(X)$ such that $\{W(A_i)\}_{i \in I}$ has the finite intersection property. Then $\bigcap_{i \in I} W(A_i) \neq \emptyset$.*

Proof Let $\{i_k\}_{k=1}^n \subseteq I$ be any finite subset. Then, by the properties of the ultra-topology,

$$W\left(\bigcap_{k=1}^n A_{i_k}\right) = \mathrm{Ultra}(Z(X)) \setminus U\left(\bigcap_{k=1}^n A_{i_k}\right) = \mathrm{Ultra}(Z(X)) \setminus \bigcup_{k=1}^n U(A_{i_k})$$

$$= \bigcap_{k=1}^n \left(\mathrm{Ultra}(Z(X)) \setminus U(A_{i_k})\right) = \bigcap_{k=1}^n W(A_{i_k}) \neq \emptyset,$$

since $\{W(A_i)\}_{i \in I}$ has the finite intersection property. Hence by Lemma 6.17, we have $\bigcap_{k=1}^n A_{i_k} \neq \emptyset$. Since the above holds for any finite subset $\{i_k\}_{k=1}^n \subseteq I$, we see that

$$F := \{B \in Z(X) \mid \exists \{i_k\}_{k=1}^n \subseteq I \text{ s.t. } \bigcap_{k=1}^n A_{i_k} \subseteq B\}$$

is a filter. Hence there exists an ultrafilter G on $Z(X)$ such that $F \subseteq G$, by Lemma 6.7. Then for all $i \in I$, $A_i \in F \subseteq G$, so $G \in W(A_i)$ for all $i \in I$, i.e. $\bigcap_{i \in I} W(A_i) \neq \emptyset$. $\qquad\square$

Proposition 6.19 *Suppose X is a topological space. Then* Ultra$(Z(X))$ *is compact.*

Proof Suppose that $\{C_i\}_{i \in I}$ is a family of closed sets in Ultra$(Z(X))$ that has the finite intersection property. Now let $i \in I$. Since C_i is closed, $D_i := $ Ultra$(Z(X)) \backslash C_i$ is open, so $D_i = \bigcup_{j \in J_i} U(A_j)$ for some subset $\{A_j\}_{j \in J_i} \subseteq Z(X)$. Then, clearly, we have that $C_i = \bigcap_{j \in J_i} W(A_j)$.

Now define $J = \bigcup_{i \in I} J_i$ and suppose that $\{j_k\}_{k=1}^n$ is a finite subset. Then for every $k \in \{1, \ldots, n\}$ there is a $i_k \in I$ such that $j_k \in J_{i_k}$. Hence

$$\emptyset \neq \bigcap_{k=1}^n C_{i_k} = \bigcap_{k=1}^n \bigcap_{j \in J_{i_k}} W(A_j) \subseteq \bigcap_{k=1}^n W(A_{j_k})m$$

since $\{C_i\}_{i \in I}$ has the finite intersection property.

Therefore, $\{W_{A_j}\}_{j \in J}$ has the finite intersection property. Hence by Lemma 6.18, $\bigcap_{j \in J} W(A_j) \neq \emptyset$. Therefore,

$$\bigcap_{i \in I} C_i = \bigcap_{i \in I} \bigcap_{j \in J_i} W(A_j) = \bigcap_{j \in J} W(A_j) \neq \emptyset,$$

so Ultra$(Z(X))$ is compact, since $\{C_i\}_{i \in I}$ was an arbitrary family of closed sets in Ultra$(Z(X))$. $\qquad\square$

6.5 Convergence of Ultrafilters for Tychonoff Spaces

For a topological space X, we have now endowed Ultra$(Z(X))$ with a topology such that it is a compact Hausdorff space. It is now time to discuss the construction of the map $S : X \to$ Ultra$(Z(X))$. For this, we consider the notion of *convergence*. Note that we use the notation $\mathcal{N}_x$ for the set of neighbourhoods of a given point $x \in X$.

Definition 6.20 Suppose X is a topological space, F is a filter on $Z(X)$ and $x \in X$. We say that F **converges to** x if $\mathcal{N}_x \cap Z(X) \subseteq F$.

The following lemma is trivial.

Lemma 6.21 *Suppose X is a topological space and let $x \in X$. Then $\mathcal{N}_x \cap Z(X)$ is a filter on $Z(X)$.*

Combined with Lemma 6.7 applied to $Z(X)$, this has the following corollary.

Corollary 6.22 *Suppose X is a topological space and let $x \in X$. Then there is an ultrafilter F on $Z(X)$ that converges to x.*

Although the above result is useful, it does not say anything about uniqueness. First of all, an ultrafilter might converge to multiple points and secondly, there might be multiple ultrafilters converging to the same point. However, for *Tychonoff spaces* both these 'degeneracies' do not exist.

Definition 6.23 Suppose X is a topological space. Then X is called a **Tychonoff space** if it is T_1 and it satisfies the following property: for every closed $C \subseteq X$ and $x \in X \setminus C$, there is a continuous function $f : X \to [0, 1]$ such that $f(x) = 0$ and $f|_C = 1$.

Tychonoff spaces are also called *completely regular* or $T_{3.5}$. Note that discrete spaces are definitely Tychonoff. The following lemma is key in proving that the degeneracies we described above do not occur for Tychonoff spaces.

Lemma 6.24 *Suppose X is a Tychonoff space and let $x \in X$. Then, for any $A \in \mathcal{N}_x$, there is a $B \in \mathcal{N}_x \cap Z(X)$ such that $B \subseteq A$.*

Proof Since $A \in \mathcal{N}_x$, there is an open set $U \in X$ such that $x \in U \subseteq A$. Then $X \setminus U$ is closed in X and $x \notin X \setminus U$. Hence there is a continuous function $f : X \to [0, 1]$ such that $f(x) = 1$ and $f|_{X \setminus U} = 0$. Then define $g := \max\{\frac{1}{2} - f, 0\}$. Then certainly, $g : X \to [0, 1]$ is continuous and

$$g^{-1}(\{0\}) = \{z \in X \mid f(z) \geq \frac{1}{2}\} \subseteq U.$$

Furthermore, $U' := \{z \in X \mid f(z) > \frac{1}{2}\}$ is open and $x \in U' \subseteq g^{-1}(\{0\}) \subseteq U$. Hence we have $g^{-1}(\{0\}) \in \mathcal{N}_x \cap Z(X)$ and $g^{-1}(\{0\}) \subseteq A$.

Using this, we can prove the first 'non-degeneracy'.

Proposition 6.25 *Suppose X is a Tychonoff space. Then every filter on $Z(X)$ converges to at most one point.*

Proof Suppose F is a filter on $Z(X)$ such that F converges to both $x, y \in X$. Then $\mathcal{N}_x \cap Z(X) \in F$ and $\mathcal{N}_y \cap Z(X) \in F$.

If $x \neq y$, there are open $U, V \subseteq X$ such that $U \cap V = \emptyset$, $x \in U$ and $y \in V$, since X is a Tychonoff space and hence Hausdorff. Since $U \in \mathcal{N}_x$ and $V \in \mathcal{N}_y$, Lemma 6.24 then gives us $A \in \mathcal{N}_x \cap Z(X)$ and $B \in \mathcal{N}_y \cap Z(X)$ such that $A \subseteq U$ and $B \subseteq V$. However, by assumption, then $A, B \in F$, so $\emptyset = A \cap B \in F$. This is a contradiction, so $x = y$.

Hence F converges to at most one point.   □

Next, for a Tychonoff space, we can describe the ultrafilters that converge to a certain point explicitly.

Proposition 6.26 *Suppose X is a Tychonoff space and let $x \in X$. Furthermore, suppose that F is an ultrafilter on $Z(X)$ that converges to x. Then*

$$F = \{A \in Z(X) \mid x \in A\}.$$

Proof Suppose there is an $A \in F$ such that $x \notin A$. Then $A \in Z(X)$, so certainly $A \subseteq X$ is closed. Hence $X \setminus A \in \mathcal{N}_x$. So, by Lemma 6.24, there is a $B \in \mathcal{N}_x \cap Z(X)$ such that $B \subseteq X \setminus A$. However, since F converges to x, then $B \in F$, and hence $\emptyset = A \cap B \in F$. This is a contradiction. Hence $x \in A$ for all $A \in F$, i.e.

$$F \subseteq \{A \in Z(X) \mid x \in A\}.$$

Next, suppose $C \in Z(X)$ such that $x \in C$. By the above, $x \in A$ for all $A \in F$, so for all $A \in F$, $A \cap C \neq \emptyset$. Hence $C \in F$, by Lemma 6.6. Hence, indeed,

$$F = \{A \in Z(X) \mid x \in A\}. \qquad \square$$

Corollary 6.22 combined with the above explicit description of a convergent ultrafilter has the following corollary.

Corollary 6.27 *Suppose X is a Tychonoff space and let $x \in X$. Then there is a unique ultrafilter on $Z(X)$ that converges to x.*

We write F_x for the unique ultrafilter that converges to the point x, i.e.

$$F_x = \{A \in Z(X) \mid x \in A\}.$$

Furthermore, we write $S : X \to \mathrm{Ultra}(Z(X))$, $x \mapsto F_x$. We will now show that this map has the desired properties.

Proposition 6.28 *Suppose X is a Tychonoff space. Then $S : X \to \mathrm{Ultra}(Z(X))$ is continuous.*

Proof Let $A \in Z(X)$ and let $x \in S^{-1}(U(A))$. Then $F_x \in U(A)$, so $A \notin F_x$, i.e. $x \notin A$, by Proposition 6.26. Since $A \in Z(X)$, A is closed, whence $X \setminus A$ is an open neighbourhood of x.

Now, let $y \in X \setminus A$. Then $y \notin A$, so $A \notin F_y$, by Proposition 6.26. Therefore, we know that $S(y) \in U(A)$, i.e. $y \in S^{-1}(U(A))$. Hence $X \setminus A \subseteq S^{-1}(U(A))$. Since $U(A)$ is an arbitrary base element of $\mathrm{Ultra}(Z(X))$, we conclude that S is continuous. $\qquad \square$

In order to prove the universal property of the Stone-Čech compactification, we extend the above result.

Proposition 6.29 *Suppose X is a Tychonoff space. Then $S : X \to \mathrm{Ultra}(Z(X))$ is an embedding.*

Proof By Proposition 6.25, we know that S is injective, whence $S : X \to S(X)$ is a bijection. Furthermore, according to Proposition 6.28, S is continuous, so we only need to prove that the function $S : X \to S(X)$ is open.

For this, let $V \subseteq X$ be open, and let $F \in S(V)$. Then there is an $x \in V$ such that $F = F_x$. Since $X \setminus V$ is closed, $x \notin X \setminus V$ and X is Tychonoff, there is a continuous function $f : X \to [0, 1]$ such that $f(x) = 1$ and $f|_{X \setminus V} = 0$.

Now, define $B := f^{-1}(\{0\}) \in Z(X)$. Then $X \setminus V \subseteq B$ and $x \notin B$. Then $B \notin F_x$, so $F_x \in U(B)$.

Let $G \in U(B) \cap S(X)$, i.e. let $y \in X$ such that $F_y \in U(B)$. Then $B \notin F_y$, so $y \notin B$ by Proposition 6.26. Then also $y \notin X \setminus V$, i.e. $y \in V$. Therefore, $U(B) \cap S(X) \subseteq S(V)$, i.e. $S : X \to S(X)$ is open.

Hence S is indeed an embedding.   □

Not only is S an embedding, its image is also dense.

Lemma 6.30 *Suppose X is a Tychonoff space. Then $S(X) \subseteq \text{Ultra}(Z(X))$ is dense.*

Proof Suppose $A \in Z(X)$ such that $U(A) \neq \emptyset$, i.e. such that $A \neq X$. Then, we have $X \setminus A \neq \emptyset$, so there is a $x \in X \setminus A$. Then $x \notin A$. Therefore, $A \notin F_x$, i.e. $F_x \in U(A)$. Hence we have that $U(A) \cap S(X) \neq \emptyset$. Since $U(A)$ was an arbitrary base element, we conclude that $S(X) \subseteq \text{Ultra}(Z(X))$ is dense.   □

6.6   Pushforward

We have now constructed the pair $(\text{Ultra}(Z(X)), S)$ for all Tychonoff spaces X. To prove that this pair in fact has the universal property of the Stone-Čech compactification, we consider *pushforwards* of filters.

Definition 6.31 Suppose X and Y are topological spaces, F is a filter on $Z(X)$ and $f : X \to Y$ is a continuous function. Then the **pushforward** of F over f is defined as

$$f_*(F) = \{A \in Z(Y) \mid f^{-1}(A) \in F\}.$$

Lemma 6.32 *Suppose X and Y are topological spaces, F is a filter on $Z(X)$ and $f : X \to Y$ is a continuous function. Then $f_*(F)$ is a filter on $Z(Y)$. Moreover, if F is an ultrafilter, then $f_*(F)$ is an ultrafilter too.*

Proof Since taking pre-images behaves nicely with respect to intersections, unions and subsets, it is clear that $f_*(F)$ is a filter.

Now, also assume that F is an ultrafilter. Then suppose $A, B \in Z(Y)$ such that $A \cup B \in f_*(F)$. Then we know that $f^{-1}(A) \cup f^{-1}(B) = f^{-1}(A \cup B) \in F$ and also that $f^{-1}(A), f^{-1}(B) \in Z(X)$. Since F is an ultrafilter, it is prime (by Lemma 6.10), so either $f^{-1}(A) \in F$ or $f^{-1}(B) \in F$. Therefore, $A \in f_*(F)$ or $B \in f_*(F)$, i.e. F is prime and hence an ultrafilter.   □

Now for any continuous function $f : X \to Y$ between any two topological spaces X and Y, consider $f_* : \text{Ultra}(Z(X)) \to \text{Ultra}(Z(Y))$ as a function.

Lemma 6.33 *Suppose X and Y are topological spaces and let $f : X \to Y$ be continuous. Then f_* is continuous.*

Proof Suppose $A \in Z(Y)$ and let $F \in f_*^{-1}(U(A))$. Then $A \notin f_*(F)$, which means that $f^{-1}(A) \notin F$, i.e. we have $F \in U(f^{-1}(A))$.

Let $G \in U(f^{-1}(A))$. Then $f^{-1}(A) \notin G$, so $A \notin f_*(G)$, i.e. $f_*(G) \in U(A)$, so $G \in f_*^{-1}(U(A))$. Hence $U(f^{-1}(A)) \subseteq f_*^{-1}(U(A))$, i.e. f_* is continuous. □

6.7 Convergence of Ultrafilters for Compact Hausdorff Spaces

We have already discussed the notion of convergence for Tychonoff spaces. For compact Hausdorff spaces, we can prove even stronger statements. First of all, since we have the following result for compact spaces.

Lemma 6.34 *Suppose X is a compact space. Then every $F \in \mathrm{Ultra}(Z(X))$ converges to at least one point.*

Proof Suppose $F \in \mathrm{Ultra}(Z(X))$ is an ultrafilter that converges to no point. Then, for all $x \in X$, we know that there is an $A_x \in \mathcal{N}_x \cap Z(X)$ such that $A_x \notin F$. Then for all $x \in X$, there is an open $V_x \subseteq X$ such that $x \in V_x \subseteq A_x$.

Then $\bigcup_{x \in X} V_x = X$, so by compactness of X, there is a finite subset $\{x_1, \dots, x_n\}$ of X such that $\bigcup_{i=1}^{n} V_{x_i} = X$. Then also $\bigcup_{i=1}^{n} A_{x_i} = X \in F$. However, F is prime, so there must be an $i \in \{1, \dots, n\}$ such that $A_{x_i} \in F$, which contradicts our assumptions.

Hence F must converge to at least one point. □

If we now also assume the Hausdorff property, this lemma has the following corollary.

Corollary 6.35 *Let K be a compact Hausdorff space. Then every $F \in \mathrm{Ultra}(Z(K))$ converges to exactly one point.*

Proof Suppose $F \in \mathrm{Ultra}(Z(K))$. Then, since K is compact, F converges to at least one point. However, since every compact Hausdorff space is T_4 and hence Tychonoff, F converges to at most one point, by Proposition 6.25. Hence F converges to exactly one point. □

For a compact Hausdorff space K, denote $\varphi_K : \mathrm{Ultra}(Z(K)) \to K$ for the unique map such that F converges to $\varphi_K(F)$ for all $F \in \mathrm{Ultra}(Z(K))$.

Proposition 6.36 *Suppose K is a compact Hausdorff space. Then φ_K is continuous.*

Proof Suppose $V \subseteq K$ is open and let $F \in \varphi_K^{-1}(V)$. Then $\varphi_K(F) \in V$. Furthermore, $K \setminus V \subseteq K$ is closed and $\varphi_K(F) \notin K \setminus V$. Since K is Tychonoff, there is a continuous $F : K \to [0, 1]$ such that $f|_{K \setminus V} = 0$ and $f(\varphi_K(F)) = 1$.

Now define $B := f^{-1}(\{0\}) \in Z(K)$ and note that $\varphi_K(F) \notin B$, so $B \notin F$, i.e. we have $F \in U(B)$. Now let $G \in U(B)$. Then $B \notin G$, so $\varphi_K(G) \notin B$. Since $K \setminus V \subseteq B$, then $\varphi_K(G) \notin K \setminus V$, so $\varphi_K(G) \in V$.

Therefore, $\varphi_K(U(B)) \subseteq V$, i.e. $U(B) \subseteq \varphi_K^{-1}(V)$. Hence $\varphi_K^{-1}(V)$ is open, i.e. φ_K is continuous. □

6.8 Universal Property

For any Tychonoff space X, compact Hausdorff space K and continuous $f : X \to K$, we now have the following diagram:

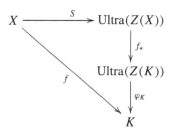

We first show that this diagram commutes.

Proposition 6.37 *Suppose X is a Tychonoff space, K a compact Hausdorff space and $f : X \to K$ a continuous function. Then $\varphi_K \circ f_* \circ S = f$.*

Proof Let $x \in X$. Then $S(x) = F_x = \{A \in Z(X) \mid x \in A\}$. Therefore,

$$
\begin{aligned}
(f_* \circ S)(x) = f_*(F_x) &= \{A \in Z(Y) \mid f^{-1}(A) \in F_x\} \\
&= \{A \in Z(Y) \mid x \in f^{-1}(A)\} \\
&= \{A \in Z(Y) \mid f(x) \in A\} \\
&= F_{f(x)}.
\end{aligned}
$$

Therefore, $(\varphi_K \circ f_* \circ S)(x) = \varphi_K(F_{f(x)}) = f(x)$, i.e. $\varphi_K \circ f_* \circ S = f$.   □

Using this, we come to the main point.

Corollary 6.38 *Suppose X is a Tychonoff space. Then $\mathrm{Ultra}(Z(X))$ together with the function $S : X \to \mathrm{Ultra}(Z(X))$, defined by $S(x) = \{A \in Z(X) \mid x \in A\}$, is the Stone-Čech compactification of X.*

Proof We showed that $\mathrm{Ultra}(Z(X))$ is a compact Hausdorff space (Propositions 6.16 and 6.19) and that S is continuous (Proposition 6.28). Therefore, we only need to show that the pair $(\mathrm{Ultra}(Z(X)), S)$ has the universal property. For this, let K be a compact Hausdorff space and $f : X \to K$ a continuous function. Then define $\beta f = \varphi_K \circ f_*$. By Proposition 6.37, then $\beta f \circ S = f$.

Furthermore, since $S(X) \subseteq \mathrm{Ultra}(Z(X))$ is dense by Lemma 6.30, βf is the unique continuous function $g : \mathrm{Ultra}(Z(X)) \to K$ such that $g \circ S = f$. Hence the pair $(\mathrm{Ultra}(Z(X)), S)$ has the universal property.   □

Chapter 7
The Continuous Subalgebra and the Kadison-Singer Conjecture

The main goal of this chapter is to prove that the continuous subalgebra does not have the Kadison-Singer property. We do this in Sect. 7.4, by considering the so-called *Anderson operator*. The sections before that one provide tools for proving properties of the Anderson operator.

In Sects. 7.1 and 7.2, we construct the *Haar states*, using the results of Chap. 6. Section 7.3 contains rather technical results, which culminate in Corollary 7.17. This corollary is in fact the only thing we need from Sect. 7.3 to prove in Sect. 7.4 that the continuous subalgebra does not have the Kadison-Singer property.

In the remainder of the chapter, we prove that this implies that the mixed subalgebra does not have the Kadison-Singer property either and hence that only maximal abelian subalgebras that are unitarily equivalent to the discrete subalgebra can possibly have the Kadison-Singer property. Once we have proven this result, we are in a position to formulate the Kadison-Singer conjecture and appreciate its consequence for the classification of subalgebras with the Kadison-Singer property.

7.1 Total Sets of States

We have already seen that the set $S(A)$ of states on a C*-algebra A is a convex, compact Hausdorff space. We are now interested in special subsets of this space; so-called *total sets of states*. For this, recall that for a self-adjoint element a and a state f, $f(a)$ is real.

Definition 7.1 Suppose A is a C*-algebra and let $T \subseteq S(A)$. We say that T is a **total set of states** for A if for any self-adjoint $a = a^* \in A$ the condition $f(a) \geq 0$ for every $f \in T$ implies that $a \geq 0$.

© The Author(s) 2016
M. Stevens, *The Kadison-Singer Property*,
SpringerBriefs in Mathematical Physics 14, DOI 10.1007/978-3-319-47702-2_7

The following lemma is trivial.

Lemma 7.2 *Suppose A is a C^*-algebra and suppose $T \subseteq T' \subseteq S(A)$, where T is a total set of states for A. Then T' is a total set of states for A.*

Total sets of states have an important property.

Lemma 7.3 *Suppose A is a C^*-algebra and $T \subseteq S(A)$ is a total sets of states. Furthermore, suppose that $a = a^* \in A$, $\alpha \in \mathbb{R}$ and that $g(a) \geq \alpha$ for every $g \in T$. Then $f(a) \geq \alpha$ for all $f \in S(A)$.*

Proof Note that $a - \alpha 1$ is self-adjoint and that $g(a - \alpha 1) = g(a) - \alpha \geq 0$ for every $g \in T$. Therefore, $a - \alpha 1 \geq 0$, since T is total.

Hence for any $f \in S(A)$, $f(a - \alpha 1) \geq 0$, since f is positive. Therefore, we obtain that $f(a) - \alpha \geq 0$, i.e. $f(a) \geq \alpha$, for all $f \in S(A)$.   □

The state space $S(A)$ of a C^*-algebra A is topologized by the weak*-topology, which is generated by considering single elements $a \in A$. Since the definition of total sets of vectors concerns only self-adjoint elements, we are especially interested in those elements. Recall that the weak*-topology on the state space $S(A)$ of a C^*-algebra A is given by the subbase that consists of the elements

$$B(f, a, \varepsilon) = \{g \in S(A) : |f(a) - g(a)| < \varepsilon\},$$

where $f \in S(A)$, $a \in A$ and $\varepsilon > 0$. As it now turns out, we only have to consider those subbase elements given by self-adjoint elements.

Lemma 7.4 *Suppose A is a C^*-algebra. Then the set*

$$\{B(f, a, \varepsilon) : f \in S(A), \ a = a^* \in A, \ \varepsilon > 0\}$$

is a subbase for the weak*-topology on $S(A)$.

Proof Suppose that $a \in A$. Then $a = b + ic$, where $b, c \in A$ are self-adjoint. Hence, for $f, g \in S(A)$,

$$|f(a) - g(a)| = |f(b + ic) - g(b + ic)|$$
$$\leq |f(b) - g(b)| + |f(c) - g(c)|.$$

So, if $g \in B(f, b, \frac{\varepsilon}{2}) \cap B(f, c, \frac{\varepsilon}{2})$, then $g \in B(f, a, \varepsilon)$. Hence

$$B(f, b, \tfrac{\varepsilon}{2}) \cap B(f, c, \tfrac{\varepsilon}{2}) \subseteq B(f, a, \varepsilon).$$

Combined with the fact that $\{B(f, a, \varepsilon) : f \in S(A), a \in A, \varepsilon > 0\}$ is a subbase for the weak*-topology on $S(A)$, this shows that

$$\left[B(f, a, \varepsilon) : f \in S(A), a = a^* \in A, \varepsilon > 0 \right]$$

is a subbase too.   □

We use this fact to prove the following important lemma about total sets of states.

Lemma 7.5 *Suppose A is a C^*-algebra and $T \subseteq S(A)$ is a total set of states. Then $S(A) = \overline{\mathrm{co}(T)}$, i.e. the sets of states is the weak*-closure of the convex hull of T.*

Proof Since $T \subseteq S(A)$ and $S(A)$ is a weak*-closed convex set, $\overline{\mathrm{co}(T)} \subseteq S(A)$. Hence we only have to prove that $S(A) \subseteq \overline{\mathrm{co}(T)}$.

To see this, let $f \in S(A)$, and suppose that $f \in \bigcap_{i=1}^n B(f_i, a_i, \varepsilon_i)$, for certain states $f_i \in S(A)$, $a_i = a_i^* \in A$ and $\varepsilon_i > 0$. Since $f \in B(f_i, a_i, \varepsilon_i)$ for any $i \in \{1, \ldots, n\}$, there are $\delta_i > 0$ for all $i \in \{1, \ldots, n\}$ such that we have the inclusion $B(f, a_i, \delta_i) \subseteq B(f_i, a_i, \varepsilon_i)$ for all $i \in \{1, \ldots, n\}$.

Now define the map

$$\varphi : S(A) \to \mathbb{R}^n, \ f \mapsto (f(a_1), \ldots, f(a_n)),$$

and define $\Omega = \varphi(T)$. We claim that for every $f \in S(A)$, we have $\varphi(f) \in \overline{\mathrm{co}(\Omega)}$.

To prove this, we argue by contraposition. So suppose that $f \in S(A)$ such that $\varphi(f) \notin \overline{\mathrm{co}(\Omega)}$. Then, using a very standard result in convexity theory, we obtain a $n - 1$-dimensional hyperplane V through $\varphi(f)$ that does not intersect $\mathrm{co}(\Omega)$ and an $\alpha > 0$ such that for every $x \in V$ and $y \in \mathrm{co}(\Omega)$, $|x - y| \geq \alpha$, i.e. $\mathrm{co}(\Omega)$ is completely on one side of V and is seperated from V by a distance of at least α. Considering the normal vector n on V, this means that for any $y \in \mathrm{co}(\Omega)$, $\langle y - \varphi(f), n \rangle \geq \alpha$, where $\langle , \rangle$ is the standard inner product on $\mathbb{R}^n$.

Now write $n = (t_1, \ldots, t_n) \in \mathbb{R}^n$ and let $g \in T$. Then $\varphi(g) \in \mathrm{co}(\Omega)$, so

$$\langle (g(a_1) - f(a_1), \ldots, g(a_n) - f(a_n)), (t_1, \ldots, t_n) \rangle \geq \alpha.$$

Writing this out, one obtains $\sum_{i=1}^n t_i(g(a_i) - f(a_i)) \geq \alpha$, i.e.

$$g\left(\sum_{i=1}^n t_i a_i\right) \geq f\left(\sum_{i=1}^n t_i a_i\right) + \alpha.$$

However, $g \in T$ was arbitrary, $\sum_{i=1}^n t_i a_i$ is self-adjoint and $f \in S(A)$, so by Lemma 7.3, $f(\sum_{i=1}^n t_i a_i) \geq f(\sum_{i=1}^n t_i a_i) + \alpha$. This is a contradiction, so $\varphi(f) \in \overline{\mathrm{co}(\Omega)}$.

Now define $\delta := \min_{i \in \{1, \ldots, n\}} \delta_i$. Then $\delta > 0$, so there is an $h \in \mathrm{co}(\Omega)$ such that we have $|h - \varphi(f)| < \delta$. Since $h \in \mathrm{co}(\Omega)$, there are $\{g_i\}_{i=1}^m \subseteq T$ and $\{s_i\}_{i=1}^m \subseteq [0, 1]$ such that $h = \sum_{i=1}^m s_i \varphi(g_i)$ and $\sum_{i=1}^m s_i = 1$.

Now define $k = \sum_{i=1}^m s_i g_i \in \mathrm{co}(T)$ and let $j \in \{1, \ldots, n\}$. Note that $\varphi(k) = h$. Then

$$|k(a_j) - f(a_j)| = |\varphi(k)_j - \varphi(f)_j| = |h_j - \varphi(f)_j| \leq |h - \varphi(f)| < \delta \leq \delta_j,$$

where we used the notation x_j for the j'th coordinate of $x = (x_1, \ldots, x_n) \in \mathbb{R}^n$.

This proves that $k \in B(f, a_j, \delta_j) \subseteq B(f_j, a_j, \varepsilon_j)$. Since $j \in \{1, \dots, n\}$ was arbitrary, $k \in \bigcap_{i=1}^{n} B(f_i, a_i, \varepsilon_i)$. However, $k \in \text{co}(T)$ too, so we obtain that $\bigcap_{i=1}^{n} B(f_i, a_i, \varepsilon_i) \cap \text{co}(T) \neq \emptyset$. Since $\bigcap_{i=1}^{n} B(f_i, a_i, \varepsilon_i)$ is an arbitrary base element around f by Lemma 7.4, $f \in \overline{\text{co}(T)}$.

Now $f \in S(A)$ was arbitrary, so $S(A) \subseteq \overline{\text{co}(T)}$ and hence $S(A) = \overline{\text{co}(T)}$, as desired. ☐

The above lemma is mainly important because of the following theorem.

Theorem 7.6 *Suppose A is a C^*-algebra and $T \subseteq S(A)$ is a total set of states. Then $\partial_e S(A) \subseteq \overline{T}$.*

Proof By Lemma 7.5, $S(A) = \overline{\text{co}(T)}$. Then by the Krein-Milman theorem, i.e. Theorem B.4, $\partial_e S(A) \subseteq \overline{T}$. ☐

In the next section, we will construct a total set of states for the continuous subalgebra. Later on, we will use Theorem 7.6 for this total set to prove that the continuous subalgebra does not have the Kadison-Singer property.

7.2 Haar States

The total set of states on the continuous subalgebra that we will consider is induced by the so-called *Haar functions*. In order to describe these, first consider the set

$$Y := \{(i, j) \in (\mathbb{N} \cup \{0\})^2 : i < 2^j\}.$$

It is easily seen that the function

$$\psi : Y \to \mathbb{N}, (i, j) \mapsto i + 2^j$$

is a bijection. Now, for each pair $(i, j) \in Y$, we have a special notation for the interval $V(i, j) = [\frac{i}{2^j}, \frac{i+1}{2^j}] \subseteq [0, 1]$. Next, define the function

$$k : Y \to L^2(0, 1), (i, j) \mapsto (\sqrt{2})^j [\chi_{V(2i, j+1)} - \chi_{V(2i+1, j+1)}],$$

and using this, define $h : \mathbb{N} \to L^2(0, 1)$ by setting $h(1) = [1]$ and, if $n \geq 2$, defining $h(n) = (k \circ \psi^{-1})(n - 1)$. The set $h(\mathbb{N}) \subseteq L^2(0, 1)$ is the set of **Haar functions**.

This procedure gives

$$h(1) = [1],$$
$$h(2) = [\chi_{[0, 1/2]} - \chi_{[1/2, 1]}],$$
$$h(3) = \sqrt{2}[\chi_{[0, 1/4]} - \chi_{[1/4, 1/2]}],$$
$$h(4) = \sqrt{2}[\chi_{[1/2, 3/4]} - \chi_{[3/4, 1]}],$$

and so on. By mere writing out it follows that the Haar functions form an orthonormal set.

In fact, since the support of $h(n)$ becomes arbitrary small as n increases, but the supports of the functions $h(1 + 2^j), h(1 + 2^j), \ldots, h(2^{j+1})$ completely cover $[0, 1]$ for every $j \in \mathbb{N}$, one can see that the Haar functions actually form a basis for $L^2(0, 1)$. For more details about this, see [1, Theorem 1.4].

The Haar functions now induce the *Haar states*, which will form the total set of states we are looking for. To do this, define $H(n) : B(L^2(0, 1)) \to \mathbb{C}$ by setting $H(n)(b) = \langle b(h(n)), h(n) \rangle$, for every $n \in \mathbb{N}$. Clearly, every $H(n)$ is a state on $L^2(0, 1)$, since $H(n)(1) = \langle h(n), h(n) \rangle = 1$ and

$$H(n)(b^*b) = \langle (b^*b)(h(n)), h(n) \rangle = \langle b(h(n)), b(h(n)) \rangle = \|b(h(n))\|^2 \geq 0$$

for every $b \geq 0$. We consider H as a function, i.e. $H : \mathbb{N} \to S(B(L^2(0, 1)))$. The set $H(\mathbb{N})$ is the set of **Haar states**.

When restricting the Haar states to the continuous subalgebra $A_c \subseteq B(L^2(0, 1))$, we get a function $H' : \mathbb{N} \to S(A_c)$, given by $H'(n) = H(n)|_{A_c}$. We will refer to the elements of $H'(\mathbb{N})$ as **restricted Haar states**. The main point of this construction is the following theorem.

Theorem 7.7 *The set $H'(\mathbb{N})$ of restricted Haar states is a total set of states for A_c.*

Proof Suppose $a = a^* \in A_c$, but a is not positive. Then there is a real-valued measurable function $g : [0, 1] \to \mathbb{C}$, a set $D \subseteq [0, 1]$ and $b, c > 0$ such that $g \in a$, $g(x) < -b$ for all $x \in D$ and $\mu(D) = c$. Since $D \subseteq [0, 1]$ is measurable, there is an open set $U \subseteq [0, 1]$ such that $D \subseteq U$ and $\mu(U \setminus D) < \frac{bc}{2\|a\|}$.

Now note that the Haar functions satisfy $h(i + 2^j - 1)^2 = 2^j[\chi_{V(i,j)}]$ for $j \geq 1$. Hence, since the $V(i, j)$ partition $[0, 1]$ in arbitrarily small intervals, we can write $\chi_U = \sum_{n=1}^{\infty} \lambda_n h(n)^2$, for some $\lambda_n \geq 0$. Then compute:

$$\sum_{n=1}^{\infty} \lambda_n \int_{[0,1]} g(x)h(n)(x)^2 = \int_{[0,1]} g(x) \sum_{n=1}^{\infty} \lambda_n h(n)(x)^2$$

$$= \int_{[0,1]} g(x)\chi_U(x) = \int_U g(x)$$

$$= \int_D g(x) + \int_{U \setminus D} g(x)$$

$$\leq -bc + \mu(U \setminus D)\|a\|$$

$$< -bc + \frac{bc}{2} = -\frac{bc}{2} < 0.$$

Since every $\lambda_n \geq 0$, there is then at least one $n \in \mathbb{N}$ such that

$$H'(n)(M_a) = \langle M_{[g]}h(n), h(n) \rangle = \int_{[0,1]} g(x)h(n)(x)^2 < 0.$$

So, whenever a self-adjoint element $b = b^* \in A_c$ satisfies $H'(n)(M_b) \geq 0$ for every $n \in \mathbb{N}$, then $b \geq 0$, i.e. the set $H'(\mathbb{N})$ is a total set of states for A_c. □

We can now use the (restricted) Haar states in combination with the concept of the Stone-Čech compactification of $\mathbb{N}$, which is Ultra($\mathbb{N}$) according to Corollary 6.38. Since $\mathbb{N}$ is discrete, the map $H : \mathbb{N} \to S(B(L^2(0, 1)))$ is continuous. Furthermore, $S(B(L^2(0, 1)))$ is a compact Hausdorff space by Proposition 3.6, so there is a unique continuous map $\beta H : \text{Ultra}(\mathbb{N}) \to S(B(L^2(0, 1)))$ such that the following diagram commutes:

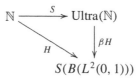

where S is the map such that $S(n) = F_n$, the principal ultrafilter belonging to $n \in \mathbb{N}$.

Likewise, for the restricted Haar states map $H' : \mathbb{N} \to S(A_c)$, there is a unique continuous map $\beta H' : \text{Ultra}(\mathbb{N}) \to S(A_c)$ such that the following diagram commutes:

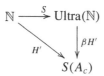

We can make a connection between these two diagrams by considering the multiplication operator $M : A_c \to B(L^2(0, 1))$, which is an inclusion. We are especially interested in the pullback of this map, i.e.

$$M^* : S(B(L^2(0, 1))) \to S(A_c),$$

given by $(M^*(f))(a) = f(M(a))$, since M^* is continuous by Lemma 3.18.

We use the map M^* for the following trivial fact.

Lemma 7.8 *The following identity holds:* $H' = M^* \circ H$.

Proof Suppose $n \in \mathbb{N}$ and $a \in A_c$. Then

$$H'(n)(a) = H(n)|_{A_c}(a) = H(n)(M(a)) = (M^*(H(n)))(a),$$

which proves that $H'(n) = (M^* \circ H)(n)$, i.e. $H' = M^* \circ H$, as desired. □

This induces the following important identity.

Corollary 7.9 *The following identity holds:* $\beta H' = M^* \circ \beta H$.

Proof First of all, note that $M^* \circ \beta H : \text{Ultra}(\mathbb{N}) \to S(A_c)$ is a continuous function, since both M^* and βH are continuous, by Lemma 3.18 and the universal property of the Stone-Čech compactifaction.

Next, note that $\beta H \circ S = H$, again by the universal property of the Stone-Čech compactification. Therefore, using Lemma 7.8,

$$M^* \circ \beta H \circ S = M^* \circ H = H'.$$

Therefore, by uniqueness of the map $\beta H'$, we have $M^* \circ \beta H = \beta H'$, as desired. $\square$

We are mainly considering the Stone-Čech compactification of $\mathbb{N}$ and the Haar states because of the following statement.

Theorem 7.10 *The following inclusion holds:* $\partial_e S(A_c) \subseteq (\beta H')(\text{Ultra}(\mathbb{N}))$.

Proof By Theorem 7.7 we know that $H'(\mathbb{N})$ is a total set of states for A_c. Then, since $H' = \beta H' \circ S$, $H'(\mathbb{N}) \subseteq (\beta H')(\text{Ultra}(\mathbb{N}))$, whence by Lemma 7.2, $(\beta H')(\text{Ultra}(\mathbb{N}))$ is a total set of states.

Therefore, by Theorem 7.6, $\partial_e S(A_c) \subseteq \overline{(\beta H')(\text{Ultra}(\mathbb{N}))}$. However, $\text{Ultra}(\mathbb{N})$ is a compact space, and $\beta H'$ is a continuous map. Therefore, $(\beta H')(\text{Ultra}(\mathbb{N})) \subseteq S(A_c)$ is compact too. Since $S(A_c)$ is Hausdorff, this implies that $(\beta H')(\text{Ultra}(\mathbb{N}))$ is closed. Therefore, $\partial_e S(A_c) \subseteq (\beta H')(\text{Ultra}(\mathbb{N}))$, as desired. $\square$

Since we are interested in the pure states on the continuous subalgebra, we are now interested in a more precise expression of the image of βH and $\beta H'$. We can describe both of them by generalizing the structure they share.

Proposition 7.11 *Suppose X is a discrete space and A a C^*-algebra. Furthermore, suppose that $F : X \to S(A)$ is some function. Then for any $a \in A$ and $U \in \text{Ultra}(X)$,*

$$\{(\beta F)(U)(a)\} = \bigcap_{\sigma \in U} \overline{\{F(x)(a) : x \in \sigma\}}.$$

Proof Let $a \in A$ and $U \in \text{Ultra}(\mathbb{N})$. Note that $\beta F = \varphi_{S(A)} \circ F_*$ by Proposition 6.37. Hence $\mathcal{N}_{(\beta F)(U)} \subseteq F_*(U)$, so for every $N \in \mathcal{N}_{(\beta F)(U)}$ we have $F^{-1}(N) \in U$.

Now let $\sigma \in U$ and let $\varepsilon > 0$. Then $B((\beta F)(U), a, \varepsilon) \in \mathcal{N}_{(\beta F)(U)}$, so

$$C_\varepsilon := \{x \in X : |F(x)(a) - (\beta F)(U)(a)| < \varepsilon\} = F^{-1}(B((\beta F)(U), a, \varepsilon)) \in U.$$

Then $\sigma, C_\varepsilon \in U$, so $\sigma \cap C_\varepsilon \in U$, i.e. there is an $x \in \sigma \cap C_\varepsilon$, i.e. there is an $x \in \sigma$ such that $|F(x)(a) - (\beta F)(U)(a)| < \varepsilon$. Therefore, we obtain that $(\beta F)(U)(a) \in \overline{\{F(x)(a) : x \in \sigma\}}$. Since $\sigma \in U$ was arbitrary, we then have

$$(\beta F)(U)(a) \in \bigcap_{\sigma \in U} \overline{\{F(x)(a) : x \in \sigma\}}.$$

Now suppose $y \in \bigcap_{\sigma \in U} \overline{\{F(x)(a) : x \in \sigma\}}$ too. Then let $\delta > 0$. Since $C_{\frac{\delta}{2}} \in U$, we have $y \in \overline{\{F(x)(a) : x \in C_{\frac{\delta}{2}}\}}$. Hence there is a $x \in C_{\frac{\delta}{2}}$ such that we obtain the inequality $|y - F(x)(a)| < \frac{\delta}{2}$. Therefore,

$$|y - (\beta F)(U)(a)| \leq |y - F(x)(a)| + |F(x)(a) - (\beta F)(x)(a)| < \frac{\delta}{2} + \frac{\delta}{2} = \delta.$$

Since $\delta > 0$ was arbitrary, $y = (\beta F)(U)(a)$. So, indeed,

$$\{(\beta F)(U)(a)\} = \bigcap_{\sigma \in U} \overline{\{F(x)(a) : x \in \sigma\}},$$

as desired.   □

7.3 Projections in the Continuous Subalgebra

We now begin with our proof of the fact that the continuous subalgebra A_c does not have the Kadison-Singer property. We first prove some rather technical results.

First of all, for a pure state $f \in \partial_e S(A_c)$, a measurable and bounded function $h : [0, 1] \to \mathbb{C}$ such that $[h] \in A_c$ is a positive element, and some $\varepsilon > 0$, we define the set

$$X(f, h, \varepsilon) := \{x \in [0, 1] : h(x) \in [f([h]) - \varepsilon, f([h]) + \epsilon]\}.$$

Note that by construction of $A_c = L^\infty(0, 1)$, for any positive $[h] \in A_c$, $f \in \partial_e S(A_c)$ and $\varepsilon > 0$, the number

$$\alpha(f, [h], \varepsilon) = \mu(X(f, h, \varepsilon))$$

is well defined, where μ is the standard measure on $[0, 1]$. In order to prove a crucial result about these numbers $\alpha(f, a, \varepsilon)$, we need to define the essential infimum of a positive element of A_c.

Definition 7.12 Suppose $a \in A_c$ is positive. Then the **essential infimum** of a is defined as

$$\text{ess inf}(a) = \inf\{t : \mu(\{x \in [0, 1] : h(x) < t\}) = 0\},$$

where $h : [0, 1] \to \mathbb{C}$ is any positive measurable function such that $[h] = a$.

Note that the essential infimum is well defined, i.e. independent of choice of representative, exactly by construction of A_c. The essential infimum has an important property when considering states.

Lemma 7.13 *Suppose $a \in A_c$ is a positive element and $f \in \partial_e S(A_c)$ is a pure state such that $f(a) = 0$. Then* ess inf$(a) = 0$.

Proof Suppose $\varepsilon > 0$ and suppose that ess inf$(a) > 0$. Then there is a positive, measurable function $h : [0, 1] \to \mathbb{C}$ such that $[h] = a$ and a $t > 0$ such that $h(x) \geq t$ for all $x \in [0, 1]$. Then $h - t1$ is still a positive, measurable function, and $[h - t1] = a - t1$. Hence $a - t1$ is a positive element of A_c, whence $-t = f(a) - tf(1) = f(a - t1) \geq 0$. This is a contradiction, so indeed, ess inf$(a) = 0$. □

Lemma 7.14 *Suppose $f \in \partial_e S(A_c)$. Let $a \in A_c$ be a positive element and let $\varepsilon > 0$. Then $\alpha(f, a, \varepsilon) \neq 0$.*

Proof Suppose $h \in a$ is a measurable, positive function. Now, consider the set

$$Z := \{x \in [0, 1] : h(x) \leq f(a)\},$$

and denote $W := [0, 1] \setminus Z$. Writing χ_Z for the characteristic function of Z, we note that $f([\chi_Z])^2 = f([\chi_Z])$, since f is multiplicative. Therefore, $f([\chi_Z]) \in \{0, 1\}$. Furthemore, $f([\chi_W]) = 1 - f([\chi_Z])$, where χ_W is the characteristic function of W. Hence, there are two cases. First, suppose that $f([\chi_Z]) = 0$. Then $f([\chi_W]) = 1$. Therefore,

$$f([\chi_W h]) = f([\chi_W])f([h]) = f(a).$$

So, writing $b = [\chi_W h] - f(a)1$, $f(b) = 0$. Furthermore, $\chi_W h - f(a)1$ is a positive function by construction, since $W = \{x \in [0, 1] : h(x) > f(a)\}$. Hence, by Lemma 7.13, ess inf$(b) = 0$, so $\mu(\{x \in [0, 1] : (\chi_W h - f(a)1)(x) < \epsilon\} \neq 0$. However,

$$\{x \in [0, 1] : (\chi_W h - f(a)1)(x) < \epsilon\} = \{x \in [0, 1] : h(x) \in (f(a), f(a) + \varepsilon)\},$$

so certainly

$$\alpha(f, a, \varepsilon) = \mu(X(f, h, \varepsilon)) = \mu(\{x \in [0, 1] : h(x) \in [f(a) - \varepsilon, f(a) + \varepsilon]\}) > 0.$$

Next, consider the case that $f([\chi_Z]) = 1$. Then a similar argument applied to the element $b = f(a)1 - [\chi_Z h]$ shows that $\alpha(f, a, \varepsilon) \neq 0$. □

Using this property, we can prove the following result about projections in the continuous subalgebra, where we use the theory in section C.2.

Lemma 7.15 *Suppose $f \in \partial_e S(A_c)$, let $a \in A_c$ be a positive element and let $\varepsilon > 0$. Then there is a projection $p \in A_c$ such that $f(p) = 1$ and $\|p(a - f(a)1)\| < \varepsilon$.*

Proof Let h be a positive measurable function such that $[h] = a$. Then write

$$Z = X(f, h, \frac{\varepsilon}{2}).$$

By Lemma 7.14, then $\mu(Z) \neq 0$. Therefore, $[\chi_Z]$ is a non-zero projection in A_c, where χ_Z is the characteristic function of Z.

By construction of Z, $|(\chi_z(h - f(a)1))(x)| < \frac{\varepsilon}{2}$ for every $x \in [0, 1]$, whence we have $\|[\chi_Z](a - f(a)1)\| < \varepsilon$.

So, now the only thing left to prove is that $f([\chi_Z]) = 1$. If $f([\chi_Z]) \neq 1$, then we have $f([\chi_Z]) = 0$, since χ_Z is a projection. Then we obtain that $f([\chi_W]) = 1$, where $W = [0, 1] \setminus Z$ and χ_W is the characteristic function of W. Then, by multiplicativity of f, $f([\chi_W h]) = f(a)$, whence $X(f, \chi_W h, \varepsilon) = \emptyset$, i.e. $\alpha(f, [\chi_W h], \varepsilon) = 0$. This contradicts Lemma 7.14. Therefore, $f([\chi_Z]) = 1$, so $[\chi_Z] \in G_f$ as required. □

Using the previous results, we can consider pure states on the continuous subalgebra that do have unique pure state extensions. First of all we have the following result.

Lemma 7.16 *Suppose that $f \in \partial_e S(A_c)$ such that $\mathrm{Ext}(f) = \{g\}$. Furthermore, let $b \in B(L^2(0, 1))$ such that $g(b^*b) = 0$. Then, for every $\varepsilon > 0$, there is a projection $p \in A_c$ such that $f(p) = 1$ and $\|bp\| < \epsilon$.*

Proof Let $\varepsilon > 0$. Since $g(b^*b) = 0$, $g(-b^*b) = 0$ and b^*b is self-adjoint, so there is a $c \in A_c$ such that $-c \leq -b^*b$ and $f(-c) + \frac{\varepsilon^2}{2} > g(-b^*b) = 0$, by Proposition 3.17. Therefore, $0 \leq b^*b \leq c$ and $f(c) < \frac{\varepsilon^2}{2}$. Since then $c \geq 0, 0 \leq f(c) < \frac{\varepsilon^2}{2}$.

Since $f \in \partial_e S(A_c)$, by Lemma 7.15 there is a projection $p \in A_c$ such that $f(p) = 1$ and $\|p(c - f(c)1)\| < \frac{\varepsilon^2}{2}$. Then we also have $0 \leq pb^*bp \leq pcp = pc$, since p is a projection in the abelian subalgebra A_c, and hence

$$\|pb^*bp\| \leq \|pc\| \leq \|p(c - f(c)1)\| + \|pf(c)\|$$

$$= \|p(c - f(c)1)\| + f(c) < 2 \cdot \frac{\varepsilon^2}{2} = \varepsilon^2.$$

However, $\|bp\|^2 = \|(bp)^*bp\| = \|pb^*bp\|$, so $\|bp\| < \varepsilon$, as desired. □

Using this, we have the following corollary.

Corollary 7.17 *Suppose that $f \in \partial_e S(A_c)$ such that $\mathrm{Ext}(f) = \{g\}$. Furthermore, let $b \in B(L^2(0, 1))$ and $\varepsilon > 0$. Then there is a projection $p \in A_c$ such that $f(p) = 1$ and $\|p(b - g(b)1)p\| < \varepsilon$.*

Proof Let $b \in B(H)$ and $\varepsilon > 0$. Since g is a pure state and $g(b - g(b)1) = 0$, by Lemma C.6 there are $c_1, c_2 \in L_g$ such that $b - g(b)1 = c_1 + c_2^*$.

Now, by Lemma 7.16, there is a projection $d_1 \in A_c$ such that $f(d_1) = 1$ and $\|c_1 d_1\| < \frac{\varepsilon}{2}$. Likewise, there is a projection $d_2 \in A_c$ such that $f(d_2) = 1$ and $\|c_2 d_2\| < \frac{\varepsilon}{2}$.

Now define $p = d_1 d_2 = d_2 d_1$. Then p is also a projection, since A_c is abelian, and $f(p) = 1$ by Lemma 4. Now we have

$$\|c_1 p\| = \|c_1 d_1 d_2\| \leq \|c_1 d_1\| < \frac{\varepsilon}{2},$$

and

$$\|pc_2^*\| = \|c_2 p\| = \|c_2 d_2 d_1\| \leq \|c_2 d_2\| < \frac{\varepsilon}{2}.$$

Therefore,

$$\|p(b - g(b)1)p\| = \|p(c_1 + c_2^*)p\| \leq \|pc_1 p\| + \|pc_2^* p\|$$
$$\leq \|c_1 p\| + \|pc_2^*\| < \frac{\varepsilon}{2} + \frac{\varepsilon}{2} = \varepsilon.$$

$\square$

We will later use the above result to show that a pure state on A_c cannot have a unique extension, which implies that A_c does not have the Kadison-Singer property.

7.4 The Anderson Operator

We are now in the position to put all the pieces of the puzzle together and prove that the continuous subalgebra does not have the Kadison-Singer property. We do this by means of the *Anderson operator*. To do this, first consider the function $\varphi : \mathbb{N} \to \mathbb{N}$, defined by:

$$\varphi(n) = \begin{cases} 1 & : n = 2 \\ n + 1 & : n \neq 2^j \; \forall j \in \mathbb{N} \\ 2^j + 1 & : \exists j \in \mathbb{N} : n = 2^{j+1}, \end{cases}$$

i.e. φ permutes 1 and 2 and all the mutual disjoint subsets $(2^j + 1, \ldots, 2^{j+1})$ (where $j = 1, 2, \ldots$) in a cyclic manner. Clearly, φ is a bijection.

Now, taking the Haar functions $h(\mathbb{N})$, the operator $\hat{U} : L^2(0, 1) \to L^2(0, 1)$ defined by $\hat{U}(h(n)) = h(\varphi(n))$ is a unitary operator, since it permutes the orthonormal basis $h(\mathbb{N})$. Furthermore, since φ has no fixed points, we obtain that $H(n)(\hat{U}) = \langle \hat{U}(h(n)), h(n) \rangle = 0$ for every $n \in \mathbb{N}$. We call $\hat{U}$ the **Anderson operator**. This operator has an important property.

Proposition 7.18 *Suppose $p \in A_c$ is a non-zero projection. Then $\|p\hat{U}p\| = 1$.*

Proof Let $\frac{1}{2} > \varepsilon > 0$. We will prove that $\|p\hat{U}p\| > 1 - \varepsilon$, and thereby conclude that $\|p\hat{U}p\| = 1$, since $\|p\hat{U}p\| \leq \|p\|^2 \|\hat{U}\| = 1$.

Now, p is non-zero, so Corollary B.27 gives us an $f \in \Omega(A_c) = \partial_e S(A_c)$ such that $f(p) = 1$. Combining Theorems 7.7 and 7.6, then $\partial_e S(A_C) \subseteq \overline{H'(\mathbb{N})}$. Therefore, $f \in \overline{H'(\mathbb{N})}$.

Since the function $g : [0, \frac{1}{2}) \to \mathbb{R}$ given by $g(t) = \sqrt{1 - 2t} - \sqrt{2t}$ is continous and satisfies $g(0) = 1$, there is a $0 < \delta < \frac{1}{2}$ such that $\sqrt{1 - 2\delta} - \sqrt{2\delta} > 1 - \varepsilon$.

Now note that $B(f, p, \delta) \subseteq S(A_c)$ is open, so $B(f, p, \delta) \cap H'(\mathbb{N}) \neq \emptyset$. Hence there is a $n \in \mathbb{N}$ such that $|f(p) - H'(n)(p)| < \delta$, i.e. $H'(n)(p) > 1 - \delta$.

Using Lemma 5.6, we see that $p = [\chi_W]$, for some measurable $W \subseteq [0, 1]$ such that $\mu(W) > 0$. Using this, and choosing a $g_n \in h(n)$,

$$1 - \delta < H'(n)(p) = \langle ph(n), h(n) \rangle = \int_{[0,1]} \chi_W g_n^2 = \int_W g_n^2.$$

Now, writing $n = i + 2^j - 1$, with $(i, j) \in Y$, we note that $h(n)^2 = 2^j [\chi_{V(i,j)}]$. Therefore,

$$1 - \delta < \int_W 2^j \chi_{V(i,j)} = 2^j \mu(W \cap V(i, j)),$$

i.e. $\mu(W \cap V(i, j)) > \frac{1}{2^j}(1 - \delta)$. Now note that we can split $V(i, j)$ into two disjoint, equal parts, i.e. $V(i, j) = V(2i, j + 1) \cup V(2i + 1, j + 1)$. Then:

$$\mu(V(i, j) \cap W) = \mu(V(2i, j + 1) \cap W) + \mu(V(2i + 1, j + 1) \cap W)$$
$$\leq \mu(V(2i, j + 1) \cap W) + \mu(V(2i + 1, j + 1)),$$

whence

$$\mu(V(2i, j + 1) \cap W) \geq \mu(V(i, j) \cap W) - \mu(V(2i + 1, j + 1))$$
$$> \frac{1}{2^j}(1 - \delta) - \frac{1}{2^{j+1}} = (\frac{1}{2})^{j+1}(1 - 2\delta).$$

Likewise, it follows that $\mu(V(2i + 1, j + 1) \cap W) > (\frac{1}{2})^{j+1}(1 - 2\delta)$. Now, upon defining $m := 2i + 2^{j+1} - 1$ we obtain that $h(m)^2 = 2^{j+1} \chi_{V(2i,j+1)}$ and also that $h(m + 1)^2 = 2^{j+1} \chi_{V(2i+1,j+1)}$. Furthermore, $\hat{U}(h(m)) = h(m + 1)$. Now choose representatives $g_m \in h(m)$ and $g_{m+1} \in h(m + 1)$. Then:

$$\|ph(m + 1)\|^2 = \langle ph(m + 1), ph(m + 1) \rangle = \int_{[0,1]} \chi_W g_{m+1}^2$$
$$= \int_W 2^{j+1} \chi_{V(2i+1,j+1)} = 2^{j+1} \mu(W \cap V(2i + 1, j + 1))$$
$$> 1 - 2\delta.$$

Furthermore, writing $Z := [0, 1] \setminus W$,

$$\|h(m) - ph(m)\|^2 = \int_{[0,1]} \chi_Z g_m^2$$
$$= \int_Z g_m^2 = \int_{[0,1]} g_m^2 - \int_W g_m^2$$
$$= 1 - \int_W \chi_{V(2i,j+1)} = 1 - \mu(W \cap V(2i, j + 1))$$
$$< 1 - (1 - 2\delta) = 2\delta.$$

Combining this, we get:

$$\|p\hat{U}ph(m)\| \geq \|p\hat{U}h(m)\| - \|p\hat{U}(h(m) - ph(m))\|$$
$$\geq \|ph(m+1)\| - \|h(m) - ph(m)\|$$
$$> \sqrt{1 - 2\delta} - \sqrt{2\delta}$$
$$> 1 - \varepsilon.$$

Since $\|h(m)\| = 1$, we then have $\|p\hat{U}p\| > 1 - \varepsilon$. Therefore, indeed, we obtain that $\|p\hat{U}p\| = 1$. $\qquad\square$

We can use this rather technical result to prove the following very important theorem.

Theorem 7.19 *Suppose $f \in \partial_e S(A_c)$. Then* Ext(f) *has more than one element.*

Proof We argue by contraposition, so we suppose Ext(f) does not have more than one element. Then by Theorem 3.16, we know that Ext(f) has exactly one element.

Now, by Theorem 7.10, there is an ultrafilter $U \in$ Ultra$(\mathbb{N})$ such that we have $(\beta H')(U) = f$. Then by Corollary 7.9, $(M^* \circ \beta H)(U) = f$. Therefore, we have $(\beta H)(U) \in$ Ext(f). Since Ext(f) consists of exactly one element, we know that Ext$(f) = \{(\beta H)(U)\}$.

Recall that the Anderson operator $\hat{U}$ satisfies $H(n)(\hat{U}) = 0$ for every $n \in \mathbb{N}$. Therefore, using Proposition 7.11,

$$\{(\beta H)(U)(\hat{U})\} = \bigcap_{\sigma \in U} \overline{\{H(n)(\hat{U}) : n \in \sigma\}} = \{0\},$$

i.e. $(\beta H)(U)(\hat{U}) = 0$.

We can now apply Corollary 7.17 to find a projection $p \in A_C$ such that $|f(p)| = 1$ and $\|p\hat{U}p\| < \frac{1}{2}$. However, then p is a non-zero projection. Therefore, by Proposition 7.18, $\|p\hat{U}p\| = 1$. This is a contradiction. Therefore, Ext(f) has more than one element, as desired. $\qquad\square$

Since A_c does have pure states by the Gelfand representation (Theorem B.25), the above theorem has the following immediate corollary, which is the result we were primarily interested in.

Corollary 7.20 *A_c does not have the Kadison-Singer property.*

7.5 The Kadison-Singer Conjecture

In the light of Corollary 5.26, the statement in Corollary 7.20 is very important. We have now eliminated the continuous subalgebra from the list of algebras that could possibly have the Kadison-Singer property. However, we can also eliminate the mixed subalgebra, by using Theorem 4.13.

Corollary 7.21 *Suppose* $1 \leq j \leq \aleph_0$. *Then* $A_d(j) \oplus A_c \subseteq B(H_j)$ *does not have the Kadison-Singer property.*

Proof Suppose $A_d(j) \oplus A_c \subseteq B(H_j)$ does have the Kadison-Singer property. Then by Theorem 4.13, $A_c \subseteq B(L^2(0, 1))$ has the Kadison-Singer property. This is in contradiction with Corollary 7.20, so $A_d(j) \oplus A_c \subseteq B(H_j)$ does not have the Kadison-Singer property. $\qquad\square$

Now that we have eliminated the continuous and mixed subalgebra of our list, we can make a new step towards our classification of abelian C*-subalgebras with the Kadison-Singer property: only the discrete subalgebra can possibly have this property. The proof of the following corollary mainly serves as a summary of our results so far.

Corollary 7.22 *Suppose H is a separable Hilbert space and $A \subseteq B(H)$ is a unital abelian C*-subalgebra that has the Kadison-Singer property. Then A is unitarily equivalent to $A_d(j) \subseteq B(\ell^2(j))$ for some $1 \leq j \leq \aleph_0$.*

Proof By Corollary 5.26, we know that A is unitarily equivalent to either $A_d(j)$, A_c or $A_d(j) \oplus A_c$ for some $1 \leq j \leq \aleph_0$. If it would be unitarily equivalent to A_c, then A_c would have the Kadison-Singer property too, by Theorem 5.3. However, this is in contradiction with Corollary 7.20. So A is not unitarily equivalent to A_c. Likewise, A is not unitarily equivalent to $A_d(j) \oplus A_c$ for some $1 \leq j \leq \aleph_0$, by Corollary 7.21. Hence we obtain that there is only one case left: A is unitarily equivalent to $A_d(j)$ for some $1 \leq j \leq \aleph_0$. $\qquad\square$

The natural question that now arises is whether we can reduce our list of abelian C*-algebras that possibly have the Kadison-Singer property even further. Note that we have already proven in Theorem 2.14 that $A_d(j)$ has the Kadison-Singer property for $j \in \mathbb{N}$. Hence the only open question is whether $A_d(\aleph_0) = \ell^\infty(\mathbb{N}) \subseteq B(\ell^2(\mathbb{N}))$ has the Kadison-Singer property. Richard Kadison and Isadore Singer [2] formulated this question in 1959 and believed that the answer was negative.

This open question became known as the **Kadison-Singer conjecture** and was answered in 2013, by Adam Marcus, Daniel Spielman and Nikhil Srivastava. Despite the belief of Kadison and Singer, it was proven that $\ell^\infty(\mathbb{N}) \subseteq B(L^2(0, 1))$ in fact *does* have the Kadison-Singer property. In the rest of this text, we will prove this and thereby conclude our classification of abelian unital C*-algebras with the Kadison-Singer property.

References

1. Wojtaszczyk, P.: A Mathematical Introduction to Wavelets. Cambridge University Press (1997)
2. Kadison, R.V., Singer, I.: Extensions of pure states. American Journal of Mathematics **81**(2), 383–400 (1959)

Chapter 8
The Kadison-Singer Problem

In the previous chapter, we have reduced the classification of unital abelian C*-algebras with the Kadison-Singer property to the Kadison-Singer conjecture. In this chapter, we show that the Kadison-Singer conjecture has a positive answer, i.e. that the algebra $\ell^\infty(\mathbb{N}) \subseteq B(\ell^2(\mathbb{N}))$ has the Kadison-Singer property.

Unexpectedly, we do this by a series of statements in the field of linear algebra. This can be done by using the reduction of the Kadison-Singer conjecture via the paving [1] and Weaver [2] conjectures. These reductions were established over the last decade and enabled the mathematicians Adam Marcus, Daniel Spielman and Nikhil Srivastava to finally prove the Kadison-Singer conjecture in 2013 [3].

We will first prove their two most important results, viz. Theorems 8.17 and 8.26. Using these results, we prove the Weaver conjecture and the paving conjecture. After that, the Kadison-Singer conjecture is easily solved.

8.1 Real Stable Polynomials

The results of Marcus, Spielman and Srivastava involve the notion of *real stable polynomials*. This theory has been developed by many mathematicians, for example by Borcea and Brändén in [4]. We define the **open upper half-plane** $\mathbb{H} \subseteq \mathbb{C}$ by

$$\mathbb{H} := \{z \in \mathbb{C} \mid \mathrm{Im}(z) > 0\},$$

and consider $\mathbb{H}^n$ as as subset of $\mathbb{C}^n$. We use this to define *real stable polynomials*.

Definition 8.1 A polynomial p in n variables is called **real stable** if all coefficients of p are real and p has no zeroes in $\mathbb{H}^n$.

© The Author(s) 2016
M. Stevens, *The Kadison-Singer Property*,
SpringerBriefs in Mathematical Physics 14, DOI 10.1007/978-3-319-47702-2_8

We first focus on real stable polynomials in one variable. We can describe these in a quite easy manner. First of all, we have the following result.

Lemma 8.2 *Suppose r is a real stable polynomial in one variable and $\{z_i\}_{i=1}^n \subseteq \mathbb{C}$ is the set of roots of r. Then $\{z_i\}_{i=1}^n \subseteq \mathbb{R}$.*

Proof Let $i \in \{1, \ldots, n\}$. Then $r(z_i) = 0$ and since r has real coefficients by real stability, we know that $r(\overline{z_i}) = \overline{r(z_i)} = 0$. Therefore, $z_i, \overline{z_i} \notin \mathbb{H}$, i.e. $Im(z_i) \leq 0$ and $-Im(z_i) = Im(\overline{z_i}) \leq 0$. Hence $Im(z_i) = 0$.

Since $i \in \{1, \ldots, n\}$ was arbitrary, we therefore have $\{z_i\}_{i=1}^n \subseteq \mathbb{R}$. □

Using this, we get the following equivalent definition of real stable polynomials in one variable.

Lemma 8.3 *Suppose p is a polynomial in one variable. Then p is real stable if and only if all coefficients and all roots of p are real.*

Proof First, suppose that p is real stable. Then by definition all coefficients of p are real. Now write $p(z) = c \prod_{i=1}^n (z - z_i)$. Then $p(z) = cz^n + \ldots$, so c is real. Now define $q := \frac{p}{c}$ and observe that q is a polynomial in one variable with real coefficients and the same roots as p, so q is real stable. Hence, by Lemma 8.2, the roots of q (i.e. the roots of p) are real. So, all coefficients and roots of p are real.

For the converse, suppose that all coefficients and roots of p are real. Then certainly p has no roots in $\mathbb{H}$. Therefore, p is real stable. □

Since all the roots of a real stable polynomial in one variable are real, we can order them. Therefore, for such a polynomial p, we can define $\rho(p)$ to be the greatest root. We then have the following result.

Proposition 8.4 *Suppose that p and q are two monic polynomials in one variable, with the same degree. Furthermore, suppose that for any $t \in [0, 1]$, the polynomial $(1 - t)p + tq$ is real stable. Then, for any $t \in [0, 1]$, there is a $s \in [0, 1]$ such that $\rho((1 - t)p + tq) = (1 - s)\rho(p) + s\rho(q)$.*

Proof First of all, for $t = 0$ we can take $s = 0$ and likewise for $t = 1$ we take $s = 1$. Hence we can assume that $0 < t < 1$. Furthermore, since p and q are interchangeable, we can assume that $\rho(p) \leq \rho(q)$.

We first prove that $\rho((1 - t)p + tq) \leq \rho(q)$. To see this, let $x > \rho(q)$. Then also $x > \rho(p)$, and hence for any $x' \geq x$, $q(x') > 0$ and $p(x') > 0$, since both p and q are monic. Therefore, $((1-t)p+tq)(x') > 0$ for all $x' \geq x$. Hence $x > \rho((1-t)p+tq)$, since $(1 - t)p + tq$ is monic too. Since $x > \rho(q)$ was arbitrary, this implies that $\rho((1 - t)p + tq) \leq \rho(q)$, as we desired to prove.

Next, suppose that $\rho((1 - t)p + tq) < \rho(p)$. We prove that this leads to a contradiction. First, note that $(1 - t)p + tq$ is monic and $\rho((1 - t)p + tq) < \rho(p)$, so it follows that $((1 - t)p + tq)(\rho(p)) > 0$, i.e. $q(\rho(p)) > 0$.

Now, for every $s \in [0, 1]$, we can write

$$(1 - s)p + sq = \prod_{i=1}^{n}(z - z_i(s)),$$

with $z_1(s) \leq z_2(s) \leq \cdots \leq z_n(s)$, where n is the degree of both q and p. Note that each $z_i : [0, 1] \rightarrow \mathbb{R}$ is a continuous function and that each z_i is real-valued too by the assumption of real stability. Furthermore, $z_n(s) = \rho((1-s)p+sq)$ for all $s \in [0, 1]$ and hence $z_n(1) = \rho(q) \geq \rho(p)$. Also, $z_n(t) = \rho((1 - t)p + tq) < \rho(p)$. Hence, by the intermediate value theorem, there is a $t' \in [t, 1]$ such that $z_n(t') = \rho(p)$. But then $((1 - t')p + t'q)(\rho(p)) = 0$, i.e. $q(\rho(p)) = 0$. This is a contradiction.

Therefore, $\rho(p) \leq \rho((1 - t)p + tq) \leq \rho(q)$, i.e. there is a $s \in [0, 1]$ such that

$$\rho((1 - t)p + tq) = (1 - s)\rho(p) + s\rho(q),$$

as we intended to prove.   □

This has the following immediate corollary.

Corollary 8.5 *Suppose $\{p_i\}_{i=1}^{n}$ is set of polynomials in one variable of the same degree, all with leading coefficient 1. Furthermore, suppose that every polynomial $p \in \mathrm{co}(\{p_i\}_{i=1}^{n})$ is real stable. Then, for any $p \in \mathrm{co}(\{p_i\}_{i=1}^{n}), \rho(p) \in \mathrm{co}(\{\rho(p_i)\}_{i=1}^{n})$.*

Now that we have covered some of the theory of real stable polynomials in one variable, it is time to give some examples of real stable polynomials in more variables. To do this, we first need the following definition.

Definition 8.6 Suppose $\{A_i\}_{i=1}^{k} \subseteq M_n(\mathbb{C})$. Then the polynomial $q(A_1, \ldots, A_k)$ in $(k + 1)$ variables defined by

$$q(A_1, \ldots, A_k)(z_0, z_1, \ldots, z_k) = \det(z_0 1 + \sum_{i=1}^{k} z_i A_i),$$

is called the **associated polynomial** of $\{A_i\}_{i=1}^{k}$.

Associated polynomials become particularly interesting for self-adjoint matrices. For this special case, we first have the following result.

Lemma 8.7 *Suppose $\{A_i\}_{i=1}^{k} \subseteq M_n(\mathbb{C})$ is a set of self-adjoint matrices. Then*

$$q(A_1, \ldots, A_k)(\overline{z_0}, \ldots, \overline{z_k}) = \overline{q(A_1, \ldots, A_k)(z_0, \ldots, z_k)}.$$

Proof This can be computed directly:

$$q(A_1, \ldots, A_k)(\overline{z_0}, \ldots, \overline{z_k}) = \det(\overline{z_0}1 + \sum_{i=1}^{k} \overline{z_i} A_i)$$

$$= \det((z_0 1 + \sum_{i=1}^{k} z_i A_i)^*)$$

$$= \overline{\det(z_0 1 + \sum_{i=1}^{k} z_i A_i)}$$

$$= \overline{q(A_1, \ldots, A_k)(z_0, \ldots, z_k)}. \qquad \square$$

Upon further refining the special case of self-adjoint matrices to positive matrices, associated polynomials become real stable, as the following proposition states.

Proposition 8.8 *Suppose $\{A_i\}_{i=1}^{k} \subseteq M_n(\mathbb{C})$ is a set of positive matrices. Then the associated polynomial $q(A_1, \ldots, A_k)$ is real stable.*

Proof Write $p = q(A_1, \ldots, A_k)$. Applying Lemma 8.7, we see that the complex conjugates of the coefficients of p are the coefficients themselves, i.e. all coefficients of p are real.

Next, suppose that $z = (z_0, \ldots, z_k)$ is a zero of p. Then, upon defining the matrix $B := z_0 1 + \sum_{i=1}^{k} z_i A_i$, we see that $\det(B) = 0$, i.e. B is not invertible, whence not injective. Therefore, there is a non-zero vector $y \in \mathbb{C}^n$ such that $By = 0$. Then we have

$$0 = \langle By, y \rangle = z_0 \|y\|^2 + \sum_{i=1}^{k} z_i \langle A_i y, y \rangle,$$

and by taking imaginary parts,

$$\text{Im}(z_0)\|y\|^2 + \sum_{i=1}^{k} \text{Im}(z_i)\langle A_i y, y \rangle = 0.$$

Now, suppose $z \in \mathbb{H}^{k+1}$. Then $\sum_{i=1}^{k} \text{Im}(z_i)\langle A_i y, y \rangle \geq 0$, since all A_i are positive. Therefore, we must have $\text{Im}(z_0)\|y\|^2 \leq 0$. Since $\text{Im}(z_0) > 0$, we then have $\|y\| = 0$, i.e. $y = 0$, which is a contradiction. Hence we must have $z \notin \mathbb{H}^{k+1}$.

Since z was an arbitrary zero of p, we have that p is real stable, as desired. $\square$

Now that we have an example of a non-trivial family of a real stable polynomials, we can discuss transformations that preserve real stability. First of all, interchanging variables obviously preserves real stability. The following transformation is less trivial.

Lemma 8.9 *Suppose $n > 1$ and let p be a real stable polynomial in n variables. Furthermore, let $t \in \mathbb{R}$ and $i \in \{1, \ldots, n\}$. Let q be the polynomial in $n-1$ variables defined by*

$$q(z_1, \ldots, z_{n-1}) = p(z_1, \ldots, z_{i-1}, t, z_i, \ldots, z_{n-1}).$$

Then q is either real stable or identically zero.

Proof Note that by our previous observation that real stability is preserved under interchanging variables we can assume that $i = n$.

It is clear that q has real coefficients, since p has real coefficients and $t \in \mathbb{R}$. Define the sequence $\{q_m\}_{m=1}^{\infty}$ of polynomials in $n - 1$ variables by

$$q_m(z_1, \ldots, z_{n-1}) = p(z_1, \ldots, z_{n-1}, t + \tfrac{i}{m}).$$

Now note that $\mathbb{H}^{n-1}$ is open and connected. Furthermore, for every compact subset $C \subseteq \mathbb{H}^{n-1}$, the sequence $\{q_m\}_{m=1}^{\infty}$ clearly converges uniformly to q. Since $t \in \mathbb{R}$ and p is real stable, each q_m has no zeroes in $\mathbb{H}^{n-1}$. Therefore, by Hurwitz's theorem (see A.17), q is either identically zero on $\mathbb{H}^{n-1}$ or has no zeroes in $\mathbb{H}^{n-1}$. In the first case, q is obviously identically zero everywhere and in the second case q is real stable. $\square$

For the next transformation that preserves real stability, we adopt the notational convention $\partial_i = \frac{\partial}{\partial z_i}$, i.e. ∂_i is the directional derivative in the i'th coordinate.

Proposition 8.10 *Suppose p is a real stable polynomial in n variables, let $t \in \mathbb{R}$, and $i \in \{1, \ldots, n\}$. Then the polynomial $(1 + t\partial_i)p$ is real stable.*

Proof Since the property of real stability is preserved under interchanging variables, it is enough to prove the claim for $i = n$.

If $t = 0$, then the result is trivial, so we can suppose $t \neq 0$. Clearly, $(1 + t\partial_n)p$ has real coefficients, so we only have to prove that $(1 + t\partial_n)p$ has no zeroes in $\mathbb{H}^n$.

We argue by contraposition, so suppose that there is a vector $(y_1, \ldots, y_n) \in \mathbb{H}^n$ such that $((1 + t\partial_n)p)(y_1, \ldots, y_n) = 0$. Then define q as the polynomial in one variable given by

$$q(z) = p(y_1, \ldots, y_{n-1}, z).$$

Since p has no zeroes in $\mathbb{H}^n$ and $(y_1, \ldots, y_{n-1}) \in \mathbb{H}^{n-1}$, q has no zeroes in $\mathbb{H}$. So, especially, $q(y_n) \neq 0$. Now write $q(z) = \alpha \prod_{i=1}^{m}(z - w_i)$, i.e. $\{w_i\}_{i=1}^{m}$ is the set of zeroes of q, counted with multiplicity. Since q has no zeroes in $\mathbb{H}$, $\mathrm{Im}(w_i) \leq 0$ for all $1 \leq i \leq m$. Now if we write q' for the derivative of q, we obtain

$$0 = ((1 + t\partial_n)p)(y_1, \ldots, y_n) = q(y_n) + tq'(y_n).$$

Since $q(y_n) \neq 0$, we then also have

$$0 = 1 + t\frac{q'(y_n)}{q(y_n)}.$$

Now, considering the explicit form of q given above, we see that

$$q'(z) = \alpha \sum_{i=1}^{m} \prod_{j \neq i} (z - w_j),$$

whence

$$\frac{q'(y_n)}{q(y_n)} = \sum_{i=1}^{m} \frac{1}{y_n - w_i} = \sum_{i=1}^{m} \frac{\overline{y_n - w_i}}{|y_n - w_i|^2}.$$

Now,

$$0 = \operatorname{Im}\left(1 + t\frac{q'(y_n)}{q(y_n)}\right) = t\sum_{i=1}^{m} \frac{\operatorname{Im}(\overline{y_n - w_i})}{|y_n - w_i|^2} = t\sum_{i=1}^{m} \frac{\operatorname{Im}(w_i) - \operatorname{Im}(z_n)}{|y_n - w_i|^2}.$$

Since $t \neq 0$, we obtain $\sum_{i=1}^{m} \frac{\operatorname{Im}(w_i) - \operatorname{Im}(z_n)}{|y_n - w_i|^2} = 0$. However, for all $i \in \{1, \ldots, m\}$, we have $\operatorname{Im}(w_i) \leq 0 < \operatorname{Im}(z_n)$, so $\sum_{i=1}^{m} \frac{\operatorname{Im}(w_i) - \operatorname{Im}(z_n)}{|y_n - w_i|^2} < 0$. This is a contradiction, so $(1 + t\partial_n)p$ has no zeroes in $\mathbb{H}^n$ and is therefore real stable, as desired. $\qquad\square$

8.2 Realizations of Random Matrices

Using the basic theory of real stable polynomials that we have established in the previous section, we can come to the first major result of Marcus, Spielman and Srivastava. They considered so-called *mixed characteristic polynomials*.

Definition 8.11 Suppose $\{A_i\}_{i=1}^{k} \subseteq M_n(\mathbb{C})$. Then the **mixed characteristic polynomial** $\mu[A_1, \ldots, A_k]$ of the set $\{A_i\}_{i=1}^{k}$ is defined as

$$\mu[A_1, \ldots, A_k](z) = \left(\prod_{i=1}^{k}(1 - \partial_i)\right) \det\left(z1 + \sum_{j=1}^{k} z_j A_k\right)\Big|_{z_1 = \cdots = z_k = 0}.$$

Mixed characteristic polynomials become interesting for positive matrices.

Proposition 8.12 *Suppose $\{A_i\}_{i=1}^{k} \subseteq M_n(\mathbb{C})$ is a set of positive matrices. Then the mixed characteristic polynomial $\mu[A_1, \ldots, A_k]$ is real stable.*

Proof By Proposition 8.8, the associated polynomial $q(A_1, \ldots, A_k)$ is real stable. Then note that

$$\mu[A_1, \ldots, A_k](z) = \left(\prod_{i=1}^{k}(1 - \partial_i)q(A_1, \ldots, A_k)(z, z_1, \ldots, z_k)\right)\Big|_{z_1 = \cdots = z_k = 0},$$

i.e. $\mu[A_1, \ldots, A_d]$ is obtained by applying both the transformation described in Proposition 8.10 and the transformation of Lemma 8.9 k times to $q(A_1, \ldots, A_d)$. Since both transformations preserve real stability, $\mu[A_1, \ldots, A_d]$ is real stable. $\square$

Now, the first major result proven by Marcus, Spielman and Srivastava (Theorem 8.17) concerns positive matrices of rank 1. We use the notation $PR_1(m)$ for the set of positive matrices of rank 1 in $M_m(\mathbb{C})$, where $m \in \mathbb{N}$.

Lemma 8.13 *Suppose $\{A_i\}_{i=1}^k \subseteq PR_1(n)$ and let $B \in M_n(\mathbb{C})$ be arbitrary. Then the polynomial p defined by*

$$p(z_1, \ldots, z_k) = \det\left(B + \sum_{i=1}^k z_i A_i\right)$$

is affine in each coordinate.

Proof Let $j \in \{1, \ldots, k\}$. Then, since we have $A_j \in PR_1(n)$, we can choose a basis $\{e_1, \ldots, e_n\}$ such that $A_j e_i = 0$ for all $i \geq 2$, i.e. with respect to this basis we have

$$A = \begin{pmatrix} a & 0 & \cdots & 0 \\ 0 & 0 & \cdots & 0 \\ \vdots & \vdots & & \vdots \\ 0 & 0 & \cdots & 0 \end{pmatrix},$$

and for fixed $\{z_i\}_{i \neq j}$, also with respect to the basis $\{e_1, \ldots, e_n\}$,

$$B + \sum_{i \neq j} z_i A_i = \begin{pmatrix} c_{11} & c_{12} & \cdots & c_{1n} \\ c_{21} & c_{22} & \cdots & c_{2n} \\ \vdots & \vdots & & \vdots \\ c_{n1} & c_{n2} & \cdots & c_{nn} \end{pmatrix},$$

for some constants $\{c_{ml}\}$. Therefore, we have

$$p(z_1, \ldots, z_k) = \det \begin{pmatrix} c_{11} + z_j a & c_{12} & \cdots & c_{1n} \\ c_{21} & c_{22} & \cdots & c_{2n} \\ \vdots & \vdots & & \vdots \\ c_{n1} & c_{n2} & \cdots & c_{nn} \end{pmatrix},$$

which has a constant and a linear term in z_j. Therefore, p is affine in z_j. $\square$

Using the above lemma, we can now see the relevance of mixed characteristic polynomials. Here, we denote the characteristic polynomial of a matrix A by p_A, i.e.

$$p_A(z) = \det(z1 - A)$$

for all $z \in \mathbb{C}$.

Lemma 8.14 *Suppose that $\{A_i\}_{i=1}^k \subseteq PR_1(n)$ and define $A = \sum_{i=1}^k A_i$. Then we have the identity $p_A = \mu[A_1, \ldots, A_k]$.*

Proof Define the polynomial p by $p(z_1, \ldots, z_k) = \det(z1 + \sum_{i=1}^k A_i)$. Then, according to Lemma 8.13, p is affine in each coordinate. Therefore, p is equal to its Taylor expansion up to order $(1, \ldots, 1)$, i.e. for any $(w_1, \ldots, w_k) \in \mathbb{C}^k$, we have

$$p(w_1, \ldots, w_k) = \Big(\sum_{j_i \in \{0,1\}} \prod_{i=1}^k t_i \partial_i^{j_i} \Big)(p(z_1, \ldots, z_k))|_{z_1 = \cdots = z_k = 0}.$$

However, $\sum_{j_i \in \{0,1\}} \prod_{i=1}^k t_i \partial_i^{j_i} = \prod_{i=1}^k (1 + t_i \partial_i)$, so

$$p(w_1, \ldots, w_k) = \Big(\prod_{i=1}^k (1 + t_i \partial_i) \Big)(p(z_1, \ldots, z_k))|_{z_1 = \cdots = z_k = 0}.$$

Now choose $w_1 = \cdots = w_k = -1$. Then:

$$
\begin{aligned}
p_A(z) &= p(w_1, \ldots, w_k) \\
&= \Big(\prod_{i=1}^k (1 + t_i \partial_i) \Big) \det(z1 + \sum_{i=1}^k A_i)|_{z_1 = \cdots = z_k = 0} \\
&= \mu[A_1, \ldots, A_k](z),
\end{aligned}
$$

i.e. $p_A = \mu[A_1, \ldots, A_k]$, as desired. $\qquad\square$

The first major result of Marcus, Spielman and Srivastava concerns random variables taking value in sets of matrices, often called random matrices. We call the outcomes of such random variables *realizations*. Furthermore, these random variables induce other random variables in a canonical way, for example by means of considering characteristic polynomials and expectation values of the original random variable. As it turns out, the statement of Lemma 8.14 behaves nicely with respect to expectation values.

Proposition 8.15 *Suppose $\{Y_i\}_{i=1}^k$ is a set of random variables taking values in $PR_1(m)$ and define $Y = \sum_{i=1}^k Y_i$. Then $\mathbb{E}p_Y = \mu[\mathbb{E}Y_1, \ldots, \mathbb{E}Y_k]$.*

Proof By Lemma 8.14, $\mathbb{E}p_Y = \mathbb{E}\mu[Y_1, \ldots, Y_k]$. Now let $B \in M_n(\mathbb{C})$ and suppose that $\{A_i\}_{i=1}^k \in M_n(\mathbb{C})$ too. Then define

$$I = \{(i_1, \ldots, i_j) \mid j \in \mathbb{N}, 1 \leq i_1 < i_2 < \cdots < i_j \leq k\},$$

and use the shorthand notation $i = (i_1, \ldots, i_j) \in I$. Then note that by Lemma 8.13 there are certain constants $\{c_i\}_{i \in I}$ such that $\det(B + \sum_{i=1}^k z_i A_i) = \sum_{i \in I} c_i z_{i_1} \cdots z_{i_j}$.

Furthermore, note that any constant c_i is given by a sum $c_i = \sum_l b_l a_{l,1} \cdots a_{l,j}$, where $i = (i_1, \ldots, i_j)$ and each $a_{l,m}$ is a coefficient of A_{i_m}.

Now, note that if we replace the set $\{A_i\}_{i=1}^k$ with the independent set of random variables $\{Y_i\}_{i=1}^k$, we obtain $\mathbb{E}c_i = \sum_l b_l \mathbb{E}a_{l,1} \cdots \mathbb{E}a_{l,j}$, since the set of random variables $\{Y_i\}_{i=1}^k$ is independent, and therefore separate coordinates are too. Hence

$$\mathbb{E} \det\Big(B + \sum_{i=1}^k z_i Y_i\Big) = \det\Big(B + \sum_{i=1}^k z_i \mathbb{E}Y_i\Big).$$

Now replace B with $z1$ and observe that

$$\mathbb{E}p_Y = \mathbb{E}\mu[Y_1, \ldots, Y_k]$$

$$= \mathbb{E}\Big(\prod_{i=1}^k (1 - \partial_i)\Big) \det\Big(z1 + \sum_{j=1}^k z_j Y_j\Big)\big|_{z_1 = \cdots = z_k = 0}$$

$$= \Big(\prod_{i=1}^k (1 - \partial_i)\Big) \mathbb{E} \det\Big(z1 + \sum_{j=1}^k z_j Y_j\Big)\big|_{z_1 = \cdots = z_k = 0}$$

$$= \Big(\prod_{i=1}^k (1 - \partial_i)\Big) \det\Big(z1 + \sum_{j=1}^k z_j \mathbb{E}Y_j\Big)\big|_{z_1 = \cdots = z_k = 0}$$

$$= \mu[\mathbb{E}Y_1, \ldots, \mathbb{E}Y_k].$$

$\square$

Recall that for a real stable polynomial p in one variable, we have introduced the notation $\rho(p)$ for the greatest root of p. The following technical statement is a key result in our discussion.

Proposition 8.16 *Suppose $\{Y_i\}_{i=1}^k$ is a set of random variables taking a finite number of values in $PR_1(m)$. Then for any $1 \leq j \leq k$ and realization $\{A_i\}_{i=1}^{j-1}$ of $\{Y_i\}_{i=1}^{j-1}$, there is a realization A_j of Y_j such that*

$$\rho(\mu[A_1, \ldots, A_j, \mathbb{E}Y_{j+1}, \ldots, \mathbb{E}Y_k]) \leq \rho(\mu[A_1, \ldots, A_{j-1}, \mathbb{E}Y_j, \ldots, \mathbb{E}Y_k]).$$

Proof Let $1 \leq j \leq k$ and suppose that $\{A_i\}_{i=1}^{j-1}$ is a realization of $\{Y_i\}_{i=1}^{j-1}$. Furthermore, suppose that $\{B_i\}_{i=1}^r$ is the set of finite values of Y_j. For each $1 \leq i \leq r$, write p_i for the probability of B_i. Now adopt the notation of the proof of Proposition 8.15 and define

$$I' = \{(i_1, \ldots, i_l) \in I \mid \exists 1 \leq q \leq l : j = i_q\}.$$

Write c_i' for the c_i that belongs to the set $(A_1, \ldots, A_{j-1}, \mathbb{E}Y_j, \ldots, \mathbb{E}Y_k)$, and for every $s \in \{1, \ldots, r\}$ $c_i(s)$ for the c_i belonging to $(A_1, \ldots, A_{j-1}, B_s, \mathbb{E}Y_{j+1}, \ldots, \mathbb{E}Y_k)$.

Then note that by linearity, for every $i \in I'$, $c_i' = \sum_{s=1}^{r} p_s c_i(s)$, and by independence of B_s, $c_i' = \sum_{s=1}^{r} p_s c_i(s)$ for every $i \in I \setminus I'$ too. Hence

$$\mu[A_1, \ldots, A_{j-1}, \mathbb{E}Y_j, \ldots, \mathbb{E}Y_k] = \sum_{s=1}^{r} p_s \mu[A_1, \ldots, A_{j-1}, B_s, \mathbb{E}Y_{j+1}, \ldots, \mathbb{E}Y_k],$$

which is a convex sum. Hence by Corollary 8.5, $\rho(\mu[A_1, \ldots, A_{j-1}, \mathbb{E}Y_j, \ldots, \mathbb{E}Y_k])$ is in the convex hull of the set $\{\rho(\mu[A_1, \ldots, A_{j-1}, B_s, \mathbb{E}Y_{j+1}, \ldots, \mathbb{E}Y_k])\}_{s=1}^{r}$. Therefore, there is a $s \in \{1, \ldots, r\}$ such that

$$\rho(\mu[A_1, \ldots, A_{j-1}, B_s, \mathbb{E}Y_{j+1}, \ldots, \mathbb{E}Y_k]) \leq \rho(\mu[A_1, \ldots, A_{j-1}, \mathbb{E}Y_j, \ldots, \mathbb{E}Y_k]).$$

Then set $A_j := B_s$ and the desired result is proven. $\qquad\square$

The first major result of Marcus, Spielman and Srivastava is now easy to prove.

Theorem 8.17 *Suppose $\{Y_i\}_{i=1}^{n}$ is a set of independent random variables taking a finite number of values in $PR_1(m)$. Then, writing $Y = \sum_{i=1}^{n} Y_i$ there is at least one realization $\{A_i\}_{i=1}^{n}$ of the set $\{Y_i\}_{i=1}^{n}$ such that $\|A\| \leq \rho(\mathbb{E}p_Y)$, where $A = \sum_{i=1}^{n} A_i$.*

Proof By applying Proposition 8.16 n times, there is a realization $\{A_i\}_{i=1}^{n}$ of $\{Y_i\}_{i=1}^{n}$ such that $\rho(\mu[A_1, \ldots, A_n]) \leq \rho(\mu[\mathbb{E}Y_1, \ldots, \mathbb{E}Y_n])$. Now, define $A = \sum_{i=1}^{n} A_i$. Then, by Proposition 8.15 we have $\mathbb{E}p_Y = \mu[\mathbb{E}Y_1, \ldots, \mathbb{E}Y_n]$ and by Lemma 8.14 we know that $p_A = \mu[A_1, \ldots, A_n]$.

Combining all this, we obtain $\|A\| = \rho(p_A) \leq \rho(\mathbb{E}p_Y)$, since A is a positive matrix. $\qquad\square$

In Sect. 8.4, we combine this theorem with the second main result of Marcus, Spielman and Srivastava, which we prove in the next section.

8.3 Orthants and Absence of Zeroes

In the first few results, we will use the notion of logarithmic derivatives.

Definition 8.18 For a differentiable function $f : \mathbb{R}^n \to \mathbb{R}$, a point $x \in \mathbb{R}^n$ such that $f(x) \neq 0$ and $i \in \{1, \ldots, n\}$, we define the i'th **logarithmic derivative** of p at the point x as $\Phi_p^i(x) = \partial_i(\log \circ p)(x) = \frac{\partial_i p(x)}{p(x)}$.

Recall that for p a real stable polynomial in one variable, we introduced the notation of $\rho(p)$ for the largest root of p. This is characterized by the fact that p has no zeroes *above* $\rho(p)$. This notion of *above* can be extended to so-called *orthants*.

Definition 8.19 Suppose $x \in \mathbb{R}^n$ for some $n \in \mathbb{N}$. Then the **orthant** Ort(x) is defined as

$$\text{Ort}(x) = \{y \in \mathbb{R}^n \mid y_i \geq x_i \forall i\}.$$

We use these two new concepts in the following result.

Lemma 8.20 *Suppose p is a real stable polynomial in two variables and let $x \in \mathbb{R}^2$ such that p has no zeroes in the orthant Ort(x). For any $n \in \mathbb{N} \cup \{0\}$, we then have the inequality*

$$(-1)^n \left(\frac{\partial^n}{\partial z_2^n} \Phi_p^1 \right)(x) \geq 0.$$

Proof First, for all $w \in \mathbb{C}$, define $q_w(z) = p(w, z)$, a polynomial in one variable. Now, $p = \sum_{i=1}^n \alpha_i z_1^{m_i} z_2^{k_i}$ for some $\{\alpha_i\}_{i=1}^n \subseteq \mathbb{R}$, $\{m_i\}_{i=1}^n \subseteq \mathbb{N}$ and $\{k_i\}_{i=1}^n \subseteq \mathbb{N}$.

Define $k = \max_{1 \leq i \leq n} k_i$ and $I = \{i \in \{1, \ldots, n\} : k_i = k\}$. Then we obtain that $\deg(q_w) = k$ if and only if $\sum_{i \in I} \alpha_i w^{m_i} \neq 0$. Since $\sum_{i \in I} \alpha_i w^{m_i}$ is just a polynomial, we see that the set $T' := \{w \in \mathbb{C} : \deg(q_w) = k\}$ is cofinite.

Now, for every $w \in T'$, $q_w(z) = c(w) \cdot \prod_{i=1}^k (z - y_i(w))$. Furthermore, since p is a polynomial, we can assume that there is a cofinite $T \subseteq T'$ such that the functions $\{y_i\}_{i=1}^k$ are holomorphic on T, by the implicit function theorem for holomorphic functions (see Theorem 7.6 in [5]). Then obviously, $T \subseteq \mathbb{C}$ is cofinite too.

Write $v_i = y_i|_{T \cap [x_1, \infty)}$. For $w \in T \cap [x_1, \infty)$, $q_w(x_2) \neq 0$, since p has no zeroes in the orthant Ort(x). Therefore, for $w \in T \cap [x_1, \infty)$, q_w is not identically zero, whence it is real stable by Lemma 8.9. Therefore, the functions $\{v_i\}_{i=1}^k$ are real-valued.

Furthermore, for $t \in T \cap [x_1, \infty)$ and $1 \leq j \leq k$ we can apply the Cauchy-Riemann equations for the function y_j at the point $(t, 0)$. In that way, we obtain

$$v_j'(t) = \lim_{h \to 0} \frac{\text{Im}(y_j(ih))}{h},$$

so if $v_j'(t) > 0$ for some $j \in \{1, \ldots, k\}$, there is a $h > 0$ such that $\text{Im}(y_j(ih)) > 0$. For this h, we then have $(ih, y_j(ih)) \in \mathbb{H}^2$, while

$$p(ih, y_j(ih)) = q_{ih}(y_j(ih)) = c(ih) \cdot \prod_{l=1}^k (y_j(ih) - y_l(ih)) = 0.$$

This is a contradiction, whence we know that the functions $\{v_i\}_{i=1}^k$ are decreasing at every point $t \in T \cap [x_1, \infty)$.

Now observe that for any $t \in T \cap [x_1, \infty)$ we have

$$\left(\frac{\partial^n}{\partial z_2^n} (\log p) \right)(t, x_2) = \left\{ \frac{\partial^n}{\partial z_2^n} ((\log q_t)(z_2)) \right\}_{z_2 = x_2}$$

$$= \left\{ \frac{\partial^n}{\partial z_2^n} (\log(c(t) \prod_{i=1}^k (z_2 - v_i(t)))) \right\}_{z_2 = x_2}$$

$$= \left\{ \frac{\partial^n}{\partial z_2^n} \left(\log(c(t)) + \sum_{i=1}^{k} \log(z_2 - v_i(t)) \right) \right\}_{z_2 = x_2}$$

$$= (-1)^{n-1} \sum_{i=1}^{k} \frac{(n-1)!}{(x_2 - v_i(t))^n}.$$

Since p has no zeroes in the orthant $\mathrm{Ort}(x)$, we conclude that $v_i(t) < x_2$ for all $i \in \{1, \ldots, k\}$ and $t \in T \cap [x_1, \infty)$. In combination with the fact that the functions $\{v_i\}_{i=1}^{k}$ are decreasing on $T \cap [x_1, \infty)$, for every $t \in T \cap [x_1 \infty)$ we obtain

$$(-1)^n \frac{\partial^n}{\partial z_2^n} \Phi_p^1(t, x_2) = (-1)^n \frac{\partial^n}{\partial z_2^n} \frac{\partial}{\partial z_1} (\log p)(t, x_2)$$

$$= (-1)^n \frac{\partial}{\partial z_1} \frac{\partial^n}{\partial z_2^n} (\log p)(t, x_2)$$

$$= -\frac{\partial}{\partial t} \sum_{i=1}^{k} \frac{(n-1)!}{(x_2 - v_i(t))^n}$$

$$\geq 0.$$

Since $T \cap [x_1, \infty) \subseteq [x_1, \infty)$ is cofinite, for any $t \in [x_1, \infty)$ the inequality

$$(-1)^n \frac{\partial^n}{\partial z_2^n} \Phi_p^1(t, x_2) \geq 0$$

holds, so it certainly holds for $t = x_1$, as desired.   □

The above result about real stable polynomials in two variables can be extended to a result about arbitrary real stable polynomials.

Lemma 8.21 *Suppose p is a real stable polynomial in n variables and let $x \in \mathbb{R}^n$ be such that p has no zeroes in the orthant $\mathrm{Ort}(x)$. Then for any $i, j \in \{1, \ldots, n\}$ and $k \in \mathbb{N} \cup \{0\}$, we have*

$$(-1)^k \left(\frac{\partial^k}{\partial z_j^k} \Phi_p^i \right) (x) \geq 0.$$

Proof If $i \neq j$, note that by renumbering we can assume that $i = 1$ and $j = 2$. Then let q be the polynomial in two variables defined by

$$q(z_1, z_2) = p(z_1, z_2, x_3, \ldots, x_n).$$

By Lemma 8.9, q is either zero or real stable. Since $p(x) \neq 0$, we know that q is not identically zero, i.e. q is real stable. Furthermore, we have

$$(-1)^k \left(\frac{\partial^k}{\partial z_2^k} \Phi_p^1 \right)(x) = (-1)^k \left(\frac{\partial^k}{\partial z_2^k} \Phi_q^1 \right)(x_1, x_2) \geq 0,$$

where we used Lemma 8.20. If $i = j$, then by renumbering, we can assume that $i = 1$. Then let r be the polynomial in one variable defined by

$$r(z) = p(z, x_2, \dots, x_n),$$

which is non-zero since $p(x) \neq 0$. Therefore, using Lemma 8.9, we know that r is real stable, so we can write

$$r(z) = c \prod_{l=1}^{m} (z - y_l),$$

where c and all y_i are real. Then we have for all $z \geq x_1$:

$$\Phi_p^i(z, x_2, \dots, x_n) = \Phi_r(z) = \frac{(\partial r)(z)}{r(z)} = \frac{c \sum_{l=1}^{m} \prod_{k \neq l} (z - y_k)}{c \prod_{l=1}^{m} (z - y_l)} = \sum_{l=1}^{m} \frac{1}{z - y_l}.$$

Therefore, we have

$$\frac{\partial^k}{\partial z^k} \Phi_p^i(x) = (-1)^k \sum_{l=1}^{m} \frac{k!}{(x_1 - y_l)^{k+1}}.$$

Since p has no zeroes in the orthant $\mathrm{Ort}(x)$, r has no zeroes in the orthant $\mathrm{Ort}(x_1)$ either, i.e. $y_l < x_1$ for all $l \in \{1, \dots, m\}$, whence indeed $(-1)^k \frac{\partial^k}{\partial z^k} \Phi_p^i(x) \geq 0$, as desired. $\qquad \square$

The next result is an easy consequence.

Corollary 8.22 *Suppose p is a real stable polynomial in n variables and let $x \in \mathbb{R}^n$ such that p has no zeroes in $\mathrm{Ort}(x)$. Furthermore, let $i, j \in \{1, \dots, n\}$. Then the function $f_{ij} : [0, \infty) \to \mathbb{R}$, given by $f_{ij}(t) = \Phi_p^i(x + te_j)$, is positive, decreasing and convex.*

Proof Let $t \in [0, \infty)$. Then, since $\mathrm{Ort}(x + te_j) \subseteq \mathrm{Ort}(x)$, we know that p has no zeroes in $\mathrm{Ort}(x + te_j)$. Hence by Lemma 8.21 we know that $f_{ij}(t) > 0$, $(\partial f_{ij})(t) < 0$ and $(\partial^2 f_{ij})(t) > 0$. Since $t \in [0, \infty)$ was arbitrary, f_{ij} is indeed positive, decreasing and convex. $\qquad \square$

The following lemma plays a key role in the proof of the second main result established by Marcus, Spielman and Srivastava.

Lemma 8.23 *Suppose p is a real stable polynomial in n variables and let $x \in \mathbb{R}^n$ be such that p has no zeroes in the orthant $\mathrm{Ort}(x)$. Furthermore, suppose that for some $i \in \{1, \dots, n\}$, there is a $C > 0$ such that $\Phi_p^i(x) + \frac{1}{C} \leq 1$. Then $(1 - \partial_i)p$ has no zeroes in $\mathrm{Ort}(x + Ce_i)$ and, for all $j \in \{1, \dots, n\}$, we have the inequality*

$$\Phi^j_{(1-\partial_i)p}(x + Ce_i) \le \Phi^j_p(x).$$

Proof Suppose $y \in \text{Ort}(x)$. Then $y = x + t$ for some $t \in \text{Ort}(0)$. Now define $w_0 = x$ and for every $j \in \{1, \ldots, n\}$, define $w_j = w_{j-1} + t_j e_j$, i.e. such that $w_n = y$. Then by Corollary 8.22,

$$\Phi^i_p(y) = \Phi^i_p(w_n) \le \Phi^i_p(w_{n-1}) \le \cdots \le \Phi^i_p(w_1) \le \Phi^i_p(w_0) = \Phi^i_p(x) < 1.$$

Hence $\Phi^i_p(y) \neq 1$, i.e. $(\partial_i p)(y) \neq p(y)$, whence $((1 - \partial_i)p)(y) \neq 0$.

Therefore, $(1 - \partial_i)p$ has no zeroes in $\text{Ort}(x)$ and therefore certainly has no zeroes in the orthant $\text{Ort}(x + Ce_i) \subseteq \text{Ort}(x)$.

Now let $j \in \{1, \ldots, n\}$. Then note that by Corollary 8.22, we know that the function $f_{ji} : [0, \infty) \to \mathbb{R}$, given by $f_{ji}(t) = \Phi^j_p(x + te_i)$ is convex, so we have the inequality

$$f_{ji}(C) \le f_{ji}(0) + C(\partial f_{ji})(C),$$

i.e.

$$\Phi^j_p(x + Ce_i) \le \Phi^j_p(x) + C(\partial_i \Phi^j_p)(x + Ce_i).$$

Rewriting this, we obtain

$$-C(\partial_i \Phi^j_p)(x + Ce_i) \le \Phi^j_p(x) - \Phi^j_p(x + Ce_i).$$

Note that for any $y \in \text{Ort}(x)$, we have

$$(\partial_j \Phi^i_p)(y) = (\partial_j \partial_i (\log \circ p))(y) = (\partial_i \partial_j (\log \circ p))(y) = (\partial_i \Phi^j_p)(y),$$

which enables us to rewrite the above inequality as

$$-C(\partial_j \Phi^i_p)(x + Ce_i) \le \Phi^j_p(x) - \Phi^j_p(x + Ce_i).$$

Since the function f_{ji} is decreasing, by Corollary 8.22 we know that

$$\Phi^i_p(x + Ce_i) \le \Phi^i_p(x) \le 1 - \frac{1}{C},$$

i.e.

$$\frac{1}{1 - \Phi^i_p(x + Ce_i)} \le C.$$

Furthermore, Lemma 8.21 gives $-(\partial_j \Phi^i_p)(x + Ce_i) \ge 0$, so we have

$$\frac{-(\partial_j \Phi^i_p)(x + Ce_i)}{1 - \Phi^i_p(x + Ce_i)} \le -C(\partial_j \Phi^i_p)(x + Ce_i).$$

Therefore, we obtain

$$\frac{-(\partial_j \Phi^i_p)(x + Ce_i)}{1 - \Phi^i_p(x + Ce_i)} \leq \Phi^j_p(x) - \Phi^j_p(x + Ce_i).$$

Next, observe that for any $y \in \mathrm{Ort}(x)$, we have $\Phi^i_p(y) = \frac{(\partial_i p)(y)}{p(y)}$, whence

$$p(y) \cdot (1 - \Phi^i_p(y)) = p(y) - (\partial_i p)(y) = ((1 - \partial_i)p)(y),$$

so we also have

$$\log(p(y)) + \log(1 - \Phi^i_p(y)) = \log(p(y) \cdot (1 - \Phi^i_p(y))) = \log(((1 - \partial_i)p)(y)).$$

Therefore,

$$\Phi^j_{(1-\partial_i)p}(y) = \Phi^j_p(y) + \frac{(\partial_j(1 - \Phi^i_p))(y)}{(1 - \Phi^i_p)(y)} = \Phi^j_p(y) - \frac{(\partial_j \Phi^i_p)(y)}{(1 - \Phi^i_p)(y)}.$$

Using this for $y = x + Ce_i$, the above inequality gives us

$$\Phi^j_{(1-\partial_i)p}(x + Ce_i) \leq \Phi^j_p(x). \qquad \square$$

This lemma can be extended to the following statement.

Corollary 8.24 *Suppose p is a real stable polynomial in n variables and let $x \in \mathbb{R}^n$ be such that p has no zeroes in the orthant $\mathrm{Ort}(x)$. Furthermore, suppose that there is a $C > 0$ such that $\Phi^i_p(x) + \frac{1}{C} \leq 1$ for all $i \in \{1, \ldots, n\}$. Then $(\prod_{i=1}^n (1 - \partial_i))p$ has no zeroes in the orthant $\mathrm{Ort}(x + w)$, where $w = (C, \ldots, C)$.*

Proof First we define $y_0 = x$ and then, inductively, we define $y_k = y_{k-1} + Ce_k$ for every $k \in \{1, \ldots, n\}$. Likewise, we define $q_0 = p$ and $q_k = (1 - \partial_k)q_{k-1}$ for every $k \in \{1, \ldots, n\}$.

We will prove by induction that for every $k \in \{0, 1, \ldots, n\}$, q_k has no zeroes in the orthant $\mathrm{Ort}(y_k)$, and that $\Phi^i_{q_k}(y_k) \leq \Phi^i_p(x)$ for all $i \in \{1, \ldots, n\}$. For this, first of all notice that the case $k = 0$ is already covered by our assumptions. Therefore, suppose we have proven the claim for some $k < n$. Then q_k has no zeroes in $\mathrm{Ort}(y_k)$ and $\Phi^i_{q_k}(y_k) \leq \Phi^i_p(x)$, for every $i \in \{1, \ldots, n\}$. Then

$$\Phi^{k+1}_{q_k}(y_k) + \frac{1}{C} \leq \Phi^{k+1}_p(x) + \frac{1}{C} \leq 1,$$

so, by Lemma 8.23, the polynomial $q_{k+1} = (1 - \partial_{k+1})q_k$ has no zeroes in the orthant $\mathrm{Ort}(y_k + Ce_{k+1}) = \mathrm{Ort}(y_{k+1})$. Furthermore, by the same lemma, for any $i \in \{1, \ldots, n\}$, we have

$$\Phi^i_{q_{k+1}}(y_{k+1}) \le \Phi^i_{q_k}(y_k) \le \Phi^i_p(x),$$

as desired. Hence we have proven our claim by induction.

In particular, $q_n = (\prod_{i=1}^n (1 - \partial_i))p$ has no zeroes in $\mathrm{Ort}(y_n) = \mathrm{Ort}(x + w)$. □

We use this result to prove the following proposition, which is a major step towards the second main result proven by Marcus, Spielman and Srivastava.

Proposition 8.25 *Suppose $\{A_i\}_{i=1}^k \subseteq M_n(\mathbb{C})$ is a set of positive matrices and let $C > 0$. Furthermore, suppose $\sum_{i=1}^k A_i = 1$ and $\mathrm{Tr}(A_i) \le C$ for all $i \in \{1, \dots, k\}$. Define $p(z_1, \dots, z_k) = \det(\sum_{i=1}^k z_i A_i)$. Then the polynomial $(\prod_{i=1}^k (1 - \partial_i))p$ does not have a zero in the orthant $\mathrm{Ort}(x)$, where $x = ((1 + \sqrt{C})^2, \dots, (1 + \sqrt{C})^2)$.*

Proof Note that $p(z_1, \dots, z_k) = q(A_1, \dots, A_k)(z_0, \dots, z_k)|_{z_0=0}$. Since the associated polynomial $q(A_1, \dots, A_k)$ is real stable according to Proposition 8.8, Lemma 8.9 now gives us that p is real stable.

Now let $t > 0$ be arbitrary and define $w_t = (t, \dots, t) \in \mathbb{R}^k$. Then, for any point $x \in \mathrm{Ort}(w_t)$, $x_i A_i \ge t A_i$ for all $i \in \{1, \dots, k\}$, so $\sum_{i=1}^k x_i A_i \ge \sum_{i=1}^k t A_i = t 1$. Therefore, for any $y \in \mathbb{R}^n$, we have

$$\langle (\sum_{i=1}^k x_i A_i)(y), y \rangle - \langle y, y \rangle \ge 0,$$

i.e. $\langle (\sum_{i=1}^k x_i A_i)(y), y \rangle \ge \|y\|^2$. Therefore, $\sum_{i=1}^k x_i A_i$ is injective, whence surjective by a dimensional argument, and hence invertible. So $p(x) = \det(\sum_{i=1}^k x_i A_i) \ne 0$, i.e. p has no zeroes in $\mathrm{Ort}(w_t)$.

Now let $i \in \{1, \dots, k\}$. Then, using Jacobi's formula for invertible matrices (Theorem A.8),

$$\begin{aligned}
\Phi^i_p(w_t) &= \frac{\frac{\partial}{\partial z_i} p(z_1, \dots, z_k)|_{z_0=\cdots=z_k=t}}{p(w_t)} \\
&= \frac{\frac{\partial}{\partial z_i} \det(\sum_{j=1}^k z_j A_j)|_{z_0=\cdots=z_k=t}}{\det(\sum_{j=1}^k t A_j)} \\
&= \mathrm{Tr}((\sum_{j=1}^k t A_j)^{-1} \cdot A_i) \\
&= \mathrm{Tr}(\frac{1}{t} A_i) \le \frac{C}{t}.
\end{aligned}$$

Now, since $t > 0$ was arbitrary, we can choose $t = C + \sqrt{C}$. Then we have established that p does not have any zeroes in $\mathrm{Ort}(w_t)$, and

$$\Phi^i_p(w_t) + \frac{1}{1 + \sqrt{C}} \le \frac{C}{C + \sqrt{C}} + \frac{1}{1 + \sqrt{C}} = \frac{C}{C + \sqrt{C}} + \frac{\sqrt{C}}{C + \sqrt{C}} = 1,$$

so by Corollary 8.24, the polynomial $(\prod_{i=1}^{k}(1 - \partial_i))p$ has no zeroes in the orthant $\mathrm{Ort}(x)$, where

$$x = (C + \sqrt{C}, \ldots, C + \sqrt{C}) + (1 + \sqrt{C}, \ldots, 1 + \sqrt{C})$$
$$= ((1 + \sqrt{C})^2, \ldots, (1 + \sqrt{C})^2),$$

as desired.   □

Now we have all the ingredients to prove the second main result of Marcus, Spielman and Srivastava.

Theorem 8.26 *Suppose $\{Y_i\}_{i=1}^{n}$ is a set of independent random variables taking values in $PR_1(m)$ and let $C > 0$. Furthermore, let $Y = \sum_{i=1}^{n} Y_i$ and suppose that $\mathbb{E}Y = 1$ and $\mathbb{E}\|Y_i\| \leq C$ for all $i \in \{1, \ldots, n\}$. Then $\rho(\mathbb{E}p_Y) \leq (1 + \sqrt{C})^2$.*

Proof According to Proposition 8.15, we have $\mathbb{E}p_Y = \mu[\mathbb{E}Y_1, \ldots, \mathbb{E}Y_n]$, i.e.

$$\mathbb{E}p_Y(z) = \left(\prod_{i=1}^{n}(1 - \partial_i)\right) \det\left(z1 + \sum_{j=1}^{n} z_j \mathbb{E}Y_j\right)\big|_{z_1 = \cdots = z_n = 0}$$

$$= \left(\prod_{i=1}^{n}(1 - \partial_i)\right) \det\left(z \sum_{l=1}^{n} \mathbb{E}Y_j + \sum_{j=1}^{n} z_j \mathbb{E}Y_j\right)\big|_{z_1 = \cdots = z_n = 0}$$

$$= \left(\prod_{i=1}^{n}(1 - \partial_i)\right) \det\left(\sum_{j=1}^{n}(z + z_j)\mathbb{E}Y_j\right)\big|_{z_1 = \cdots = z_n = 0}$$

$$= \left(\prod_{i=1}^{n}(1 - \partial_i)\right) \det\left(\sum_{j=1}^{n} z_j \mathbb{E}Y_j\right)\big|_{z_1 = \cdots = z_n = z}.$$

Now, note that for any $i \in \{1, \ldots, n\}$ and any realization A_i of Y_i, $\mathrm{Tr}(A_i) = \|A_i\|$, since A_i is a positive matrix of rank 1. Therefore, for any $i \in \{1, \ldots, n\}$, we have

$$\mathrm{Tr}(\mathbb{E}Y_i) = \mathbb{E}(\mathrm{Tr}(Y_i)) = \mathbb{E}\|Y_i\| \leq C.$$

Therefore, applying Proposition 8.25 to $\{\mathbb{E}Y_i\}_{i=1}^{n}$, we obtain that the polynomial

$$q(z_1, \ldots z_n) = \left(\prod_{i=1}^{n}(1 - \partial_i)\right) \det\left(\sum_{j=1}^{n} z_j \mathbb{E}Y_j\right)$$

has no zero in the orthant $\mathrm{Ort}(x)$, where $x = ((1 + \sqrt{C})^2, \ldots, (1 + \sqrt{C})^2)$.

Now suppose that $\rho(\mathbb{E}p_Y) > (1 + \sqrt{C})^2$. Then $y := (\rho(\mathbb{E}p_Y), \ldots, \rho(\mathbb{E}p_Y))$ is a zero of q, and $y \in \mathrm{Ort}(x)$. This is a contradiction. Therefore, $\rho(\mathbb{E}p_Y) \leq (1 + \sqrt{C})^2$, as desired.   □

The above result can be combined with the first main result of Marcus, Spielman and Srivastava (i.e. Theorem 8.17) to prove the so-called *Weaver theorem*. This will be the main goal for the next section.

8.4 Weaver's Theorem

In 2004, Nik Weaver showed that the Kadison-Singer conjecture was equivalent to a conjecture in the field of linear algebra [2], which became known as the *Weaver conjecture*. The two main results of Marcus, Spielman and Srivastava, which we gave in Theorems 8.17 and 8.26, enable us to prove this conjecture, which is why we speak of *Weaver's theorem*. We formulate it in a slightly different way from Weaver, following Terrence Tao's blog [6].

Theorem 8.27 *Suppose $k, m, n \in \mathbb{N}$ and let $C \geq 0$. Suppose $\{A_i\}_{i=1}^{k} \subseteq PR_1(n)$, such that $\|A_i\| \leq C$ for $1 \leq i \leq k$ and $\sum_{i=1}^{k} A_i = 1$. Then there exists a partition $\{Z_i\}_{i=1}^{m}$ of $\{1, \ldots, k\}$ such that for all $j \in \{1, \ldots, m\}$,*

$$\|\sum_{i \in Z_j} A_i\| \leq \left(\frac{1}{\sqrt{m}} + \sqrt{C}\right)^2.$$

Proof Let Y_i be the random variable taking values in $\{m\left(|e_j\rangle\langle e_j| \otimes A_i\right)\}_{1 \leq j \leq m}$, with all elements having a probability of $\frac{1}{m}$. Note that for every $j \in \{1, \ldots, m\}$, we have $m\left(|e_j\rangle\langle e_j| \otimes A_i\right) \in PR_1(nm)$, since if we write $m\left(|e_j\rangle\langle e_j| \otimes A_i\right) : (\mathbb{C}^n)^m \to (\mathbb{C}^n)^m$, we have

$$m\left(|e_j\rangle\langle e_j| \otimes A_i\right) = (0, \ldots, 0, mA_i, 0, \ldots, 0),$$

with mA_i on the j'th position. Hence the rank of $m\left(|e_j\rangle\langle e_j| \otimes A_i\right)$ is equal to the rank of mA_i, which is 1 by assumption.

Now note that $\{Y_i\}_{i=1}^{k}$ is a set of independent random variables and define the new random variable $Y = \sum_{i=1}^{k} Y_i$. Then we can compute:

$$\mathbb{E}Y = \sum_{i=1}^{k} \mathbb{E}Y_i = \sum_{i=1}^{k} \sum_{j=1}^{m} \frac{1}{m} m\left(|e_j\rangle\langle e_j| \otimes A_i\right)$$

$$= \left(\sum_{j=1}^{m} |e_j\rangle\langle e_j|\right) \otimes \left(\sum_{i=1}^{k} A_i\right) = 1 \otimes 1 = 1.$$

Next, note that by our previous description,

$$\|m\left(|e_j\rangle\langle e_j| \otimes A_i\right)\| = m\|A_i\| \leq mC,$$

for all $j \in \{1, \ldots, m\}$. Therefore, $\mathbb{E}\|Y_i\| \le mC$ for all $i \in \{1, \ldots, k\}$. Hence we have $\rho(\mathbb{E}p_Y) \le (1 + \sqrt{mC})^2$, by Theorem 8.26. However, by Theorem 8.17, we know that for every $i \in \{1, \ldots, k\}$, there is a $j_i \in \{1, \ldots, m\}$ such that

$$\|\sum_{i=1}^{k} m\big(|e_{j_i}\rangle\langle e_{j_i}| \otimes A_i\big)\| \le \rho(\mathbb{E}p_Y),$$

i.e. we have

$$\|\sum_{i=1}^{k} m\big(|e_{j_i}\rangle\langle e_{j_i}| \otimes A_i\big)\| \le (1 + \sqrt{mC})^2.$$

Now, for all $j \in \{1, \ldots, m\}$ define $Z_j := \{1 \le i \le k \mid j_i = j\}$. Then $\{Z_j\}_{j=1}^{m}$ is a partition of $\{1, \ldots, k\}$. Furthermore,

$$\sum_{i=1}^{k} m\big(|e_{j_i}\rangle\langle e_{j_i}| \otimes A_i\big) = (\sum_{i \in Z_1} mA_i, \ldots, \sum_{i \in Z_m} mA_i),$$

whence

$$\|\sum_{i \in Z_j} mA_i\| \le (1 + \sqrt{mC})^2,$$

for all $j \in \{1, \ldots, m\}$. Therefore,

$$\|\sum_{i \in Z_j} A_i\| \le \left(\frac{1}{\sqrt{m}} + \sqrt{C}\right)^2,$$

for all $j \in \{1, \ldots, m\}$. $\qquad\qquad\qquad\qquad\qquad\qquad\qquad\qquad\qquad\quad\square$

We can use Weaver's theorem to prove the following result.

Proposition 8.28 *Suppose that $n \in \mathbb{N}$ and let $p \in M_n(\mathbb{C})$ be a projection. Furthermore, write $\alpha = \max_{1 \le i \le n} p_{ii}$. Then, for any $m \in \mathbb{N}$, there is a set of projections $\{q_i\}_{i=1}^{m} \subseteq D_n(\mathbb{C})$ such that $\sum_{i=1}^{m} q_i = 1$ and*

$$\|q_i p q_i\| \le \left(\sqrt{\frac{1}{m}} + \sqrt{\alpha}\right)^2$$

for all $i \in \{1, \ldots, m\}$.

Proof Let $m \in \mathbb{N}$. If $m = 1$, we can take the set of projections $q_1 = 1$, which clearly satisfies the requirements. So, suppose $m \ge 2$. Define $V := p(\mathbb{C}^n)$ and let $l := \dim(V)$. Now, for every $i \in \{1, \ldots, n\}$ define $A_i \in M_n(\mathbb{C})$ by

$$A_i(x) = \langle x, p(e_i)\rangle p(e_i),$$

for all $x \in \mathbb{C}^n$, where e_i is the i'th standard basis vector of $\mathbb{C}^n$. Then, for all $x \in \mathbb{C}^n$, $\langle A_i(x), x \rangle = |\langle x, p(e_i) \rangle|^2 \geq 0$, and $A_i(x) \in \mathbb{C}p(e_i)$, so $\{A_i\}_{i=1}^n$ is a set of positive matrices of rank 1. Furthermore, for every $i \in \{1, \ldots, n\}$ and $x \in \mathbb{C}^n$, we have $\|A_i(x)\| \leq \|x\| \|p(e_i)\|^2$, while $\|A_i(p(e_i))\| = \|p(e_i)\|^3$, so $\|A_i\| = \|p(e_i)\|^2$. However,

$$\|p(e_i)\|^2 = \langle p(e_i), p(e_i) \rangle = \langle p(e_i), e_i \rangle = p_{ii} \leq \alpha,$$

so $\|A_i\| \leq \alpha$. Now note that $\mathbb{C}^n = V \oplus V^\perp$, and that for $(v, w) \in V \oplus V^\perp$, we have

$$A_i(v, w) = \langle (v, w), p(e_i) \rangle p(e_i) = \langle v, p(e_i) \rangle p(e_i),$$

i.e. $A_i = (B_i, 0) : V \oplus V^\perp \to V \oplus V^\perp$. Then $B_i : V \to V$ is linear and hence, after choosing a basis $\{\epsilon_1, \ldots, \epsilon_l\}$ for V, we can regard B_i as an element of $M_l(\mathbb{C})$. Then $\{B_i\}_{i=1}^n \subseteq M_l(\mathbb{C})$ is a set of postive matrices of rank 1, such that $\|B_i\| = \|A_i\| \leq \alpha$ and for $v \in V$,

$$\left\langle \left(\sum_{i=1}^n B_i \right) v, v \right\rangle = \sum_{i=1}^n \langle A_i v, v \rangle = \sum_{i=1}^n |\langle v, p(e_i) \rangle|^2$$

$$= \sum_{i=1}^n |\langle p(v), e_i \rangle|^2 = \sum_{i=1}^n |\langle v, e_i \rangle|^2 = \langle v, v \rangle,$$

i.e. $\sum_{i=1}^n B_i = 1 \in M_l(\mathbb{C})$. Therefore, by Theorem 8.27 there is a partition $\{Z_i\}_{i=1}^m$ of $\{1, \ldots, n\}$ such that

$$\left\| \sum_{i \in Z_j} B_i \right\| \leq \left(\sqrt{\frac{1}{m}} + \sqrt{\alpha} \right)^2$$

for all $j \in \{1, \ldots, m\}$. Now define $\{q_i\}_{i=1}^m \subseteq D_n(\mathbb{C})$ by $(q_i)_{jj} = 1$ if $j \in Z_i$ and $(q_i)_{jj} = 0$ if $j \notin Z_i$. Then, for $i \in \{1, \ldots, m\}$, note that

$$\|q_i p q_i\| = \|(q_i p)(q_i p)^*\| = \|q_i p\|^2,$$

by the C*-identity. Now, for $(v, w) \in V \oplus V^\perp$, we have

$$\|(q_i p)(v, w)\|^2 = \|q(v)\|^2 = \sum_{i \in Z_j} |\langle v, e_i \rangle|^2 = \sum_{i \in Z_j} |\langle p(v), e_i \rangle|^2 = \sum_{i \in Z_j} |\langle v, p(e_i) \rangle|^2$$

$$= \sum_{i \in Z_j} \langle A_i(v), v \rangle = \left\langle \left(\sum_{i \in Z_j} A_i \right) (v), v \right\rangle$$

$$\leq \left\| \sum_{i \in Z_j} A_i \right\| \|v\|^2 \leq \left(\sqrt{\frac{1}{m}} + \sqrt{\alpha} \right)^2 \|(v, w)\|^2,$$

i.e. $\|q_i p q_i\| \leq \left(\sqrt{\frac{1}{m}} + \sqrt{\alpha} \right)^2$, as desired. $\square$

Now that we have the result of Proposition 8.28, we are well on track to proving the Kadison-Singer conjecture, although the results we have now obtained are (merely) results in linear algebra. By means of the so-called *paving theorems* we can step up from linear algebra to functional analysis. This will be done in the next section.

8.5 Paving Theorems

In the original article on the Kadison-Singer conjecture, written by Kadison and Singer themselves, it is already pointed out that the Kadison-Singer conjecture is equivalent to a conjecture which became known as the *paving conjecture*. We prove this conjecture in three steps, which we call the *paving theorems*. These theorems are rather technical, but enable us to prove the Kadison-Singer conjecture in the next section in a simple manner.

The first theorem deals with self-adjoint matrices. Furthermore, we use the function(s) diag : $M_n(\mathbb{C}) \to D_n(\mathbb{C})$ for every $n \in \mathbb{N}$, which take the diagonal parts of matrices, i.e. $\mathrm{diag}(a)_{ij} = 0$ if $i \neq j$ and $\mathrm{diag}(a)_{ii} = a_{ii}$.

Theorem 8.29 *Suppose $\varepsilon > 0$. Then there is an $m \in \mathbb{N}$ with the following property: for every $n \in \mathbb{N}$ and self-adjoint $a \in M_n(\mathbb{C})$ such that $\|a\| \leq 1$ and $\mathrm{diag}(a) = 0$, there are projections $\{p_i\}_{i=1}^m \subseteq D_n(\mathbb{C})$ such that $\sum_{i=1}^m p_i = 1$ and $\|p_i a p_i\| \leq \varepsilon$ for all $1 \leq i \leq m$.*

Proof Note that the function $g : [0, \infty) \to [0, \infty)$, $g(x) = 2(\sqrt{x} + \sqrt{1/2})^2 - 1$ is a continuous and strictly increasing function and that $g(0) = 0$. Therefore, there is an $x_0 > 0$ such that $g(x_0) \leq \varepsilon$. Since $\frac{1}{x_0} \in (0, \infty)$, there is an $l \in \mathbb{N}$ such that $\frac{1}{x_0} \leq l$, i.e. $\frac{1}{l} \leq x_0$, whence $g(\frac{1}{l}) \leq \varepsilon$.

Now set $m = l^2$ and let $n \in \mathbb{N}$ and $a = a^* \in M_n(\mathbb{C})$ be a matrix such that $\|a\| \leq 1$ and $\mathrm{diag}(a) = 0$.

Then note that $a^2 \geq 0$, since $a^2 = a^*a$. Furthermore, $\|a^2\| = \|a^*a\| = \|a\|^2 \leq 1$. Therefore, by Lemma B.22, $1 - a^2 \geq 0$. Therefore, there is a positive $b \in M_n(\mathbb{C})$ such that $b^2 = 1 - a^2$, by Proposition B.20. Then by Proposition B.20, we know that $ab = ba$. Now define $p \in M_{2n}(\mathbb{C})$ by

$$p = \frac{1}{2}\begin{pmatrix} 1+a & b \\ b & 1-a \end{pmatrix},$$

and observe that p is self-adjoint, since both a and b are. It is easy to show that p is a projection. Therefore, we can apply Proposition 8.28 to p. Since we have assumed that $\mathrm{diag}(a) = 0$, $p_{ii} = \frac{1}{2}$ for every $i \in \{1, \dots, 2n\}$, this means that there is a set of projections $\{q_i\}_{i=1}^l \subseteq D_{2n}(\mathbb{C})$ such that $\sum_{i=1}^l q_i = 1$ and

$$\|q_i p q_i\| \leq \left(\sqrt{\frac{1}{l}} + \sqrt{\frac{1}{2}}\right)^2 = \frac{g(1/l) + 1}{2} \leq \frac{\varepsilon + 1}{2},$$

for every $i \in \{1, \ldots, l\}$.

Now define the set of projections $\{r_i\}_{i=1}^{l} \subseteq D_n(\mathbb{C})$ by defining $(r_i)_{jj} = (q_i)_{jj}$ for every $j \in \{1, \ldots, n\}$ and likewise define the projections $\{s_i\}_{i=1}^{l} \subseteq D_n(\mathbb{C})$ by setting $(s_i)_{jj} = (q_i)_{(j+d)(j+d)}$ for every $j \in \{1, \ldots, n\}$.

Since $\sum_{i=1}^{l} q_i = 1$ by construction, we then also have $\sum_{i=1}^{l} r_i = \sum_{i=1}^{l} s_i = 1$. Then, for all $i, j \in \{1, \ldots, l\}$, define $p_{ij} = r_i s_j$. Then $\{p_{ij}\} \subseteq D_n(\mathbb{C})$ is a set of m projections. We prove that this set satisfies the desired properties. First of all,

$$\sum_{i,j} p_{ij} = \sum_{i=1}^{l} r_i \sum_{j=1}^{l} s_j = \sum_{i=1}^{l} r_i = 1.$$

Next, let $i \in \{1, \ldots, l\}$ and $x \in \mathbb{C}^n$. Then observe that $(r_i(x), 0) = q_i(x, 0)$ and hence

$$p q_i(x, 0) = p(r_i(x), 0) = \left(\left(\frac{1+a}{2}\right)(r_i(x)), \left(\frac{1}{2}b\right)(r_i(x))\right),$$

so

$$(q_i p q_i)(x, 0) = \left(\left(r_i\left(\frac{1+a}{2}\right)r_i\right)(x), s_i\left(\frac{1}{2}b\right)(r_i(x))\right),$$

whence

$$\left\|\left(r_i\left(\frac{1+a}{2}\right)r_i\right)(x)\right\| \leq \|(q_i p q_i)(x, 0)\| \leq \|q_i p q_i\| \|(x, 0)\| \leq \frac{\varepsilon + 1}{2}\|x\|.$$

Therefore, $\|r_i(1 + a)r_i\| \leq \varepsilon + 1$. Likewise, we have $\|s_j(1 - a)s_j\| \leq \varepsilon + 1$ for any $j \in \{1, \ldots, l\}$.

Since $D_n(\mathbb{C})$ is abelian, we therefore also have

$$\|p_{ij}(1 + a)p_{ij}\| = \|r_i s_j(1 + a)r_i s_j\| \leq \|s_j\|^2 \|r_i(1 + a)r_i\| \leq \varepsilon + 1,$$

and likewise $\|p_{ij}(1 - a)p_{ij}\| \leq \varepsilon + 1$.

Again, let $x \in \mathbb{C}^n$ and define $b = (\varepsilon + 1)p_{ij} - p_{ij}(1 + a)p_{ij}$. Then

$$
\begin{aligned}
\langle bx, x \rangle &= (\varepsilon + 1)\langle p_{ij}x, x \rangle - \langle p_{ij}(1 + a)p_{ij}x, x \rangle \\
&= (\varepsilon + 1)\|p_{ij}x\|^2 \langle p_{ij}(1 + a)p_{ij}p_{ij}x, p_{ij}x \rangle \\
&\geq (\varepsilon + 1)\|p_{ij}x\|^2 - \|p_{ij}(1 + a)p_{ij}\| \|p_{ij}x\|^2 \\
&\geq 0,
\end{aligned}
$$

i.e. $b \geq 0$, which implies $p_{ij}(1 + a)p_{ij} \leq (\varepsilon + 1)p_{ij}$, i.e. $p_{ij}ap_{ij} \leq \varepsilon p_{ij}$.

Likewise, it follows that $p_{ij}(1-a)p_{ij} \leq (\varepsilon+1)p_{ij}$, so $-\varepsilon p_{ij} \leq p_{ij}ap_{ij}$. Therefore, we have

$$-\varepsilon p_{ij} \leq p_{ij}ap_{ij} \leq \varepsilon p_{ij},$$

so by Lemma B.21, $\|p_{ij}ap_{ij}\| \leq \|\varepsilon p_{ij}\| \leq \varepsilon$, as desired. $\qquad\square$

The above paving theorem gives a result about self-adjoint matrices. The second paving theorem drops this condition.

Theorem 8.30 *Suppose $\varepsilon > 0$. Then there is an $l \in \mathbb{N}$ with the following property: for each $n \in \mathbb{N}$ and $a \in M_n(\mathbb{C})$ such that $\mathrm{diag}(a)=0$, there is a set of projections $\{r_i\}_{i=1}^l \subseteq D_n(\mathbb{C})$ such that $\sum_{i=1}^m r_i = 1$ and $\|r_i a r_i\| \leq \varepsilon \|a\|$.*

Proof Since $\varepsilon > 0$, by Theorem 8.29, there is an $m \in \mathbb{N}$ with the following property: for every $n \in \mathbb{N}$ and self-adjoint $a \in M_n(\mathbb{C})$ such that $\|a\| \leq 1$ and $\mathrm{diag}(a)=0$, there are projections $\{p_i\}_{i=1}^m \subseteq D_n(\mathbb{C})$ such that $\sum_{i=1}^m p_i = 1$ and $\|p_i a p_i\| \leq \varepsilon$ for all $1 \leq i \leq m$.

Now, define $l = m^2$ and let $n \in \mathbb{N}$ and $\varepsilon > 0$. If $a = 0$, then taking $r_1 = 1$ and $r_i = 0$ for all $i \in \{2,\dots,l\}$ yields the required set of projections $\{r_i\}_{i=1}^l$.

Hence, assume that $a \neq 0$. Observe that $b = \frac{a+a^*}{2}$ and $c = \frac{a-a^*}{2i}$ are self-adjoint elements of $M_n(\mathbb{C})$ and that $a = b + ic$. Furthermore, $\|b\| \leq \|a\|$ and $\|c\| \leq \|a\|$ by the triangle inequality, and $\mathrm{diag}(b) = \mathrm{diag}(c) = 0$.

Therefore, there are projections $\{p_i\}_{i=1}^m \subseteq D_n(\mathbb{C})$ and $\{q_j\}_{j=1}^m \subseteq D_n(\mathbb{C})$ such that

$$\sum_{i=1}^m p_i = 1,$$

$$\sum_{j=1}^m q_j = 1,$$

$$\left\| p_i \frac{b}{\|a\|} p_i \right\| \leq \frac{\varepsilon}{2}$$

for all $\{i \in 1,\dots,m\}$ and

$$\left\| q_j \frac{c}{\|a\|} q_j \right\| \leq \frac{\varepsilon}{2}$$

for all $j \in \{1,\dots,m\}$. Therefore, $\|p_i b p_i\| \leq \frac{\varepsilon}{2}\|a\|$ for all $i \in \{1,\dots,m\}$ and also $\|q_j c q_j\| \leq \frac{\varepsilon}{2}\|a\|$ for all $j \in \{1,\dots,m\}$.

Now, for $i,j \in \{1,\dots,m\}$, define $r_{ij} = p_i q_j \in D_n(\mathbb{C})$. Since $D_n(\mathbb{C})$ is abelian, we know that $r_{ij} = p_i q_j = q_j p_i$ is again a projection for each pair (i,j). Now note that

$$\sum_{i,j} r_{ij} = \sum_{ij} p_i q_j = \sum_{i=1}^m p_i \left(\sum_{j=1}^m q_j\right) = \sum_{i=1}^m p_i = 1,$$

and that for any pair (i, j),

$$\|r_{ij}br_{ij}\| = \|p_iq_jbp_iq_j\| = \|q_jp_ibp_iq_j\| \le \|p_ibp_i\| \le \frac{\varepsilon}{2}\|a\|,$$

and likewise $\|r_{ij}cr_{ij}\| \le \frac{\varepsilon}{2}\|a\|$. Therefore,

$$\|r_{ij}ar_{ij}\| = \|r_{ij}(b + ic)r_{ij}\| \le \|r_{ij}br_{ij}\| + \|r_{ij}cr_{ij}\| \le \frac{\varepsilon}{2}\|a\| + \frac{\varepsilon}{2}\|a\| = \varepsilon\|a\|,$$

which means that the set $\{r_{ij}\}$ satisfies all the requirements.   □

So far, we have only proven results in finite dimension in this chapter. However, the independence of $n \in \mathbb{N}$ in the second paving theorem enables us to actually prove a similar result where we replace $M_n(\mathbb{C})$ with $B(\ell^2(\mathbb{N}))$. This is the third paving theorem. For this result, we use the map diag : $B(\ell^2(\mathbb{N})) \to \ell^\infty(\mathbb{N})$ defined by $\text{diag}(a)(n) = \langle \delta_n, a\delta_n \rangle$.

Theorem 8.31 *Suppose $\varepsilon > 0$. Then there is an $l \in \mathbb{N}$ with the following property: for all $a \in B(\ell^2(\mathbb{N}))$ such that $\text{diag}(a) = 0$, there is a set of projections $\{p_i\}_{i=1}^l \subseteq \ell^\infty(\mathbb{N})$ such that $\sum_{i=1}^l p_i = 1$ and $\|p_iap_i\| \le \varepsilon\|a\|$ for every $i \in \{1, \ldots, l\}$.*

Proof By Theorem 8.30, there is an $l \in \mathbb{N}$ with the following property: if $n \in \mathbb{N}$ and $c \in M_n(\mathbb{C})$ such that $\text{diag}(c) = 0$, then there is a set of projections $\{r_i\}_{i=1}^l \subseteq D_n(\mathbb{C})$ such that $\sum_{i=1}^m r_i = 1$ and $\|r_icr_i\| \le \varepsilon\|c\|$.

Now let $a \in B(\ell^2(\mathbb{N}))$ such that $\text{diag}(a) = 0$. Then, for $n \in \mathbb{N}$, consider the function $\varphi_n : B(\ell^2(\mathbb{N})) \to M_n(\mathbb{C})$, given by $(\varphi_n(b))_{ij} = \langle b\delta_j, \delta_i \rangle$ for every $b \in B(\ell^2(\mathbb{N}))$. Then clearly $\|\varphi_n\| = 1$. Furthermore, we also have $\text{diag}(\varphi_n(a)) = 0$, since we have assumed that $\text{diag}(a) = 0$. Therefore, there is a set of projections $\{r_{n,i}\}_{i=1}^l \subseteq D_n(\mathbb{C})$ such that $\sum_{i=1}^l r_{n,i} = 1$ and

$$\|r_{n,i}\varphi_n(a)r_{n,i}\| \le \varepsilon\|\varphi_n(a)\| \le \varepsilon\|a\|,$$

for all $1 \le i \le l$.

For any fixed $i \in \{1, \ldots, l\}$, we have $\langle r_{n,i}\delta_m, \delta_m \rangle \in \{0, 1\}$ for all $m \le n$, since $r_{n,i}$ is a projection. We now prove that there is a strictly increasing function $\psi : \mathbb{N} \to \mathbb{N}$ as well as a set $\{y_i\}_{i=1}^l \subseteq \{0, 1\}^\mathbb{N}$ such that for every $1 \le i \le l$, y_i is the limit of the sequence $\{x_{i,n}\}_{n\in\mathbb{N}} \subseteq \{0, 1\}^\mathbb{N}$, where

$$x_{i,n}(m) = \begin{cases} \langle r_{\psi(n),i}\delta_m, \delta_m \rangle : m \le \psi(n) \\ 0 \qquad\qquad\qquad\quad : m > \psi(n). \end{cases}$$

We prove this by induction in l. For $l = 0$, we can simply take $\psi = \text{Id}$. Now suppose we have proven the claim for $l - 1$, i.e. there is a strictly increasing function $\psi' : \mathbb{N} \to \mathbb{N}$ and a set $\{y_i\}_{i=1}^{l-1}$ such that for every $i \in \{1, \ldots, l - 1\}$, y_i is the limit of the sequence $\{z_{i,n}\}_{n\in\mathbb{N}} \subseteq \{0, 1\}^\mathbb{N}$, where

$$z_{i,n}(m) = \begin{cases} \langle r_{\psi'(n),i}\delta_m, \delta_m \rangle & : m \le \psi'(n) \\ 0 & : m > \psi'(n). \end{cases}$$

Now define $w_n \in \{0, 1\}^{\mathbb{N}}$ by

$$w_n(m) = \begin{cases} \langle r_{\psi'(n),l}\delta_m, \delta_m \rangle & : m \le \psi'(n) \\ 0 & : m > \psi'(n). \end{cases}$$

Now, note that $\{w_n\}_{n \in \mathbb{N}}$ is a sequence in $\{0, 1\}^{\mathbb{N}}$. Furthermore, by Tychonoff's theorem (see Theorem A.12), $\{0, 1\}^{\mathbb{N}}$ is compact and by Theorem A.16 $\{0, 1\}^{\mathbb{N}}$ is also metrizable. Hence $\{w_n\}_{n \in \mathbb{N}}$ is a sequence in the compact metrizable space $\{0, 1\}^{\mathbb{N}}$ and it therefore has a subsequence $\{w_{n_k}\}_{k \in \mathbb{N}}$ that converges to some $y_l \in \{0, 1\}^{\mathbb{N}}$. The function $\varphi : \mathbb{N} \to \mathbb{N}$ defined by $\varphi(k) = n_k$ is strictly increasing and therefore the function $\psi := \varphi \circ \psi'$ is strictly increasing, too.

Now, for $i \in \{1, \dots, l\}$, define $x_{i,n} := z_{i,\varphi(n)}$. Since $\{x_{i,n}\}_{n \in \mathbb{N}}$ is then a subsequence of $\{z_{i,\varphi(n)}\}_{n \in \mathbb{N}}$, $\{x_{i,n}\}_{n \in \mathbb{N}}$ converges to y_i and satisfies

$$x_{i,n}(m) = \begin{cases} \langle r_{\psi(n),i}\delta_m, \delta_m \rangle & : m \le \psi(n) \\ 0 & : m > \psi(n) \end{cases}$$

Furthermore, define $x_{l,n} := w_{\varphi(n)}$. Then by construction, $\{x_{l,n}\}_{n \in \mathbb{N}}$ converges to y_l and is given by

$$x_{l,n}(m) = \begin{cases} \langle r_{\psi(n),l}\delta_m, \delta_m \rangle & : m \le \psi(n) \\ 0 & : m > \psi(n) \end{cases}$$

This concludes the induction step.

Now for all $i \in \{1, \dots, l\}$ define $p_i \in \ell^\infty(\mathbb{N})$ by $p_i(m) = y_i(m)$ and note that every p_i is a projection. We first prove that $\sum_{i=1}^{l} p_i = 1$. To see this, let $m \in \mathbb{N}$ and observe that for every $i \in \{1, \dots, l\}$ there is an N_i such that $x_{i,n}(m) = y_i(m)$ for every $n \ge N_i$, since $\{0, 1\}$ is discrete. Then define $N := \max_{1 \le i \le l} N_i$. Then we have

$$\sum_{i=1}^{l} p_i(m) = \sum_{i=1}^{l} y_i(m) = \sum_{i=1}^{l} x_{i,N}(m) = \sum_{i=1}^{l} \langle r_{\psi(N),i}\delta_m, \delta_m \rangle$$

$$= \langle \left(\sum_{i=1}^{l} r_{\psi(N),i} \right) \delta_m, \delta_m \rangle = \langle \delta_m, \delta_m \rangle = 1.$$

Since $m \in \mathbb{N}$ was arbitrary, $\sum_{i=1}^{l} p_i = 1$, as desired.

Now, suppose that $\psi_1, \psi_2 \in \ell^2(\mathbb{N})$ have finite support, i.e. there are $M_1, M_2 \in \mathbb{N}$ such that $\psi_1(n) = 0$ for every $n \ge M_1$ and $\psi_2(n) = 0$ for every $n \ge M_2$.

Define $M = \max(M_1, M_2)$. Then for every $m \in \{1, \dots, M\}$, there is an $N_m \in \mathbb{N}$ such that $x_{i,n}(m) = y_i(m) = p_i(m)$ for all $n \ge N_m$. Now define $N' := \max_{1 \le m \le M} N_m$ and $N := \max(N', M)$. Then consider the canonical map

$\alpha_N : \ell^2(\mathbb{N}) \to \mathbb{C}^N$ given by $(\alpha_N(h))(n) = h(n)$. Then by construction, for any $i \in \{1, \ldots, l\}$,

$$\langle p_i a p_i \psi_1, \psi_2 \rangle = \langle a p_i \psi_1, p_i \psi_2 \rangle = \langle \varphi_N(a) \alpha_N(p_i \psi_1), \alpha_N(p_i \psi_2) \rangle,$$

since the support of $p_i \psi_1$ is a subset of the support of ψ_1 and likewise for ψ_2. However, by construction of N, we have that $\alpha_N(p_i \psi_1) = r_{N,i} \psi_1$ and also that $\alpha_N(p_i \psi_2) = r_{N,i} \psi_2$. Therefore,

$$\begin{aligned}
\langle p_i a p_i \psi_1, \psi_2 \rangle &= \langle \varphi_N(a) \alpha_N(p_i \psi_1), \alpha_N(p_i \psi_2) \rangle \\
&= \langle \varphi_N(a) r_{N,i} \alpha_N(\psi_1), r_{N,i} \alpha_N(\psi_2) \rangle \\
&= \langle r_{N,i} \varphi_N(a) r_{N,i} \alpha_N(\psi_1), \alpha_N(\psi_2) \rangle \\
&\leq \| r_{N,i} \varphi_N(a) r_{N,i} \alpha_N(\psi_1) \| \| \alpha_N(\psi_2) \| \\
&\leq \| r_{N,i} \varphi_N(a) r_{N,i} \| \| \alpha_N(\psi_1) \| \| \alpha_N(\psi_2) \| \\
&= \| r_{N,i} \varphi_N(a) r_{N,i} \| \| \psi_1 \| \| \psi_2 \| \\
&= \varepsilon \| a \| \| \psi_1 \| \| \psi_2 \|.
\end{aligned}$$

Now note that $\psi_1, \psi_2 \in \ell^2(\mathbb{N})$ were arbitrary elements of finite support. Therefore, by Proposition B.10, $\| p_i a p_i \| \leq \varepsilon \| a \|$. Hence $\{p_i\}_{i=1}^{l} \subseteq \ell^\infty(\mathbb{N})$ satisfies all properties we desired. □

As we mentioned before, this final paving theorem is the last technical result before we can prove the Kadison-Singer conjecture. This will be done in the next section.

8.6 Proof of the Kadison-Singer Conjecture

Using the paving theorem (i.e. Theorem 8.31), we can give an explicit description of extensions of states on $\ell^\infty(\mathbb{N})$. We first need the following result.

Lemma 8.32 *Suppose* $f \in \partial_e S(\ell^\infty(\mathbb{N}))$, *let* $g \in \mathrm{Ext}(f)$ *and suppose* $a \in B(\ell^2(\mathbb{N}))$ *such that* $\mathrm{diag}(a) = 0$. *Then* $g(a) = 0$.

Proof Suppose $\varepsilon > 0$. By Theorem 8.31, we obtain a finite set of projections $\{p_i\}_{i=1}^{n} \subseteq \ell^\infty(\mathbb{N})$ such that $\sum_{i=1}^{n} p_i = 1$ and $\| p_i a p_i \| \leq \varepsilon \| a \|$ for every $i \in \{1, \ldots, n\}$.

Since $f \in \partial_e S(\ell^\infty(\mathbb{N})) = \Omega(\ell^\infty(\mathbb{N}))$, $f(p_i) \in \{0, 1\}$ for all $i \in \{1, \ldots, n\}$. Since also $\sum_{i=1}^{n} p_i = 1$, there is an $i_0 \in \{1, \ldots, n\}$ such that $f(p_{i_0}) = 1$ and $f(p_j) = 0$ if $j \neq i_0$. Since $g \in \mathrm{Ext}(f)$, we also have $g(p_{i_0}) = 1$ and $g(p_j) = 0$ for every $j \neq i_0$. Now, using the Cauchy-Schwarz inequality (see Lemma 3.3), we have:

$$|g(p_i a p_j)|^2 \leq g(p_i p_i^*) g((a p_j)^* a p_j) = g(p_i) g((a p_j)^* a p_j),$$

for any $i, j \in \{1, \ldots, n\}$. Hence, for $i \neq i_0$, $g(p_i a p_j) = 0$. Likewise, if $j \neq i_0$, $g(p_i a p_j) = 0$. Therefore, we can compute:

$$|g(a)| = |g\left(\left(\sum_i p_i\right) a \left(\sum_j p_j\right)\right)| = |\sum_{i,j} g(p_i a p_j)| = |g(p_{i_0} a p_{i_0})| \leq \|p_{i_0} a p_{i_0}\| \leq \varepsilon.$$

Since $\varepsilon > 0$ was arbitrary, we hence have $g(a) = 0$, as desired. □

Now, we can easily describe the extensions of states on $\ell^\infty(\mathbb{N})$.

Corollary 8.33 *Suppose $f \in \partial_e S(\ell^\infty(\mathbb{N}))$ and $g \in \mathrm{Ext}(f)$. Then $g = f \circ \mathrm{diag}$.*

Proof Suppose $a \in B(\ell^2(\mathbb{N}))$. Then $\mathrm{diag}(a - \mathrm{diag}(a)) = \mathrm{diag}(a) - \mathrm{diag}(a) = 0$, so by Lemma 8.32, we have $g(a - \mathrm{diag}(a)) = 0$, i.e. $g(a) = g(\mathrm{diag}(a)) = f(\mathrm{diag}(a))$, since $\mathrm{diag}(a) \in \ell^\infty(\mathbb{N})$. Therefore, $g = f \circ \mathrm{diag}$, as desired. □

The Kadison-Singer conjecture is now an easy corollary.

Corollary 8.34 *The subalgebra $\ell^\infty(\mathbb{N}) \subseteq B(\ell^2(\mathbb{N}))$ has the Kadison-Singer property.*

Proof Suppose that $f \in \partial_e S(\ell^\infty(\mathbb{N}))$. By Theorem 3.16, we know that $\mathrm{Ext}(f) \neq \emptyset$. Now suppose $g, h \in \mathrm{Ext}(f)$. Then by Corollary 8.33, $g = f \circ \mathrm{diag} = h$. Hence $\mathrm{Ext}(f)$ contains exactly one element. Therefore, $\ell^\infty(\mathbb{N}) \subseteq B(\ell^2(\mathbb{N}))$ has the Kadison-Singer property. □

Now that we have established the answer to the Kadison-Singer conjecture we are able to finish our classification of abelian unital C*-subalgebras with the Kadison-Singer property, in the case of a separable Hilbert space. The proof of the following is statement merely serves as a summary of the most important results of the text.

Corollary 8.35 *Suppose H is a separable Hilbert space and $A \subseteq B(H)$ is an abelian, unital C*-subalgebra. Then A has the Kadison-Singer property if and only if it is unitarily equivalent to $A_d(j)$ for some $1 \leq j \leq \aleph_0$.*

Proof In Corollary 7.22 we already established that if A has the Kadison-Singer property, then A is unitarily equivalent to $A_d(j)$ for some $1 \leq j \leq \aleph_0$.

Furthermore, for $j \in \mathbb{N}$, we showed in Theorem 2.14 that $A_d(j)$ has the Kadison-Singer property. Likewise, for $j = \aleph_0$, Corollary 8.34 shows that $A_d(j)$ has the Kadison-Singer property. Combined with Theorem 5.3, we conclude that if A is unitarily equivalent to $A_d(j)$ for some $1 \leq j \leq \aleph_0$, A has the Kadison-Singer property. □

References

1. Anderson, J.: Extensions, restrictions and representations of states on C*-algebras. Trans. Am. Math. Soc. **249**(2), 303–329 (1979)
2. Weaver, N.: The Kadison-Singer problem in discrepancy theory. Discrete Math. **278**, 227–239 (2004)
3. Marcus, A., Spielman, D., Srivastava, N.: Interlacing families II: Mixed characteristic polynomials and the Kadison-Singer problem. Ann. Math. **182**, 327–350 (2015)
4. Borcea, J., Brändén, P.: The Lee-Yang and Pólya-Schur programs. I. Linear operators preserving stability. Invent. Math. **177**(3), 541–569 (2009)
5. Fritzsche, K., Grauert, J.: From Holomorphic Functions to Complex Manifolds. Springer (2002)
6. Tao, T.: Real stable polynomials and the Kadison-Singer problem (2013). https://terrytao.wordpress.com/2013/11/04/real-stable-polynomials-and-the-kadison-singer-problem/

Appendix A
Preliminaries

Throughout the main text, we need results from a wide range of mathematics. In this appendix we discuss the required results from linear algebra, order theory, topology and complex analysis. In the next appendix we give results from functional analysis and operator algebras. Lastly, in Appendix C, we treat some results that rely on the definitions and results in the main text, but are not included in the main text itself. Together, these three appendices form the background of the main text. Most results are non-trivial, but are so general that a complete discussion (including all proofs) is beyond the scope of this text. In the case of missing proofs, we refer to some standard textbooks.

A.1 Linear Algebra

We need results from linear algebra for two main reasons. First of all, some results in functional analysis can be reduced to linear algebra. Secondly, in Chap. 8, we reduced the proof of the Kadison-Singer conjecture to results in linear algebra.

Hermitian Forms

We first concern ourselves with hermitian forms.

Lemma A.1 *Suppose V is a complex vector space and let $\sigma : V^2 \to \mathbb{C}$ be a map that is anti-linear in the first argument and linear in the second argument such that $\sigma(v, v) \in \mathbb{R}$ for each $v \in V$. Then σ is hermitian, i.e. $\sigma(v, w) = \overline{\sigma(w, v)}$ for all $v, w \in V$.*

This lemma has the following immediate corollary.

Corollary A.2 *Suppose V is a complex vector space and $\sigma : V^2 \to \mathbb{C}$ is a map that is anti-linear in the first argument and linear in the second argument such that $\sigma(v, v) \geq 0$ for all $v \in V$. Then σ is a pre-inner product, i.e. a positive hermitian form.*

© The Author(s) 2016
M. Stevens, *The Kadison-Singer Property*,
SpringerBriefs in Mathematical Physics 14, DOI 10.1007/978-3-319-47702-2

Corollary A.2 is especially important because of the *Cauchy-Schwarz inequality*.

Proposition A.3 *Suppose V is a complex vector space and $\sigma : V^2 \to \mathbb{C}$ is a pre-inner product. Then the **Cauchy-Schwarz inequality** holds: for all $a, b \in V$, we have*

$$|\sigma(a, b)|^2 \leq \sigma(a, a)\sigma(b, b).$$

Adjugate Matrices and Jacobi's Formula

In the main text, we need Jacobi's formula, which deals with adjugate matrices. To introduct these properly, we first need two other definitions.

Definition A.4 For a matrix $A \in M_n(\mathbb{C})$ and $i, j \in \{1, \ldots, n\}$, we define the matrix $r(A)(i, j) \in M_{n-1}(\mathbb{C})$ by removing the ith row and jth column from A. We call $r(A)(i, j)$ the **reduced matrix** of A at position (i, j).

Definition A.5 For a matrix $A \in M_n(\mathbb{C})$ and $i, j \in \{1, \ldots, n\}$, the **cofactor** of A at position (i, j) is given by $\text{cof}(A)(i, j) = \det(r(A)(i, j))$.

Using cofactors, we can define adjugate matrices.

Definition A.6 For a matrix $A \in M_n(\mathbb{C})$, the **adjugate matrix** $\text{adj}(A) \in M_n(\mathbb{C})$ is given by $\text{adj}(A)_{ij} = (-1)^{i+j} \text{cof}(A)(j, i)$.

Lemma A.7 *Suppose $A \in M_n(\mathbb{C})$. Then the following properties hold:*

1. $\text{adj}(A) \cdot A = \det(A)I$
2. *If A is invertible, then $A^{-1} = \frac{1}{\det(A)} \text{adj}(A)$.*

Now we can introduce the main thing we need: Jacobi's formula.

Theorem A.8 (Jacobi's formula) *Suppose $A : \mathbb{R} \to M_n(\mathbb{C})$ is a differentiable function. Then **Jacobi's formula** holds:*

$$\frac{d}{dt} \det A(t) = \text{Tr}\left(\text{adj}(A(t)) \cdot \frac{d}{dt} A(t)\right).$$

Furthermore, if $A(t)$ is invertible, then

$$\frac{\frac{d}{dt} \det A(t)}{\det(A(t))} = \text{Tr}\left(A(t)^{-1} \cdot \frac{d}{dt} A(t)\right).$$

For a more detailed account of linear algebra, see [2, 6, 10] or [17].

A.2 Order Theory

For a general introduction to the theory of partially ordered sets and lattices we refer to [3]. In the main text, we need the following result, which is not among the standard results.

Proposition A.9 *Suppose F is a maximal totally ordered subset of a lattice and $F_0 \subseteq F$. Then $\vee F_0 \in F$ and $\wedge F_0 \in F$.*

Proof Let $e \in F$. Either $f \leq e$ for all $f \in F_0$ or there is a $f \in F_0$ such that $e \leq f$. In the first case, $\vee F_0 \leq e$, and in the second case $e \leq f \leq \vee F_0$. So either $e \leq \vee F_0$ or $e \geq \vee F_0$. Therefore, $F \cup \{\vee F_0\}$ is totally ordered, so by maximality of F, $\vee F_0 \in F$.

Likewise, for every $e \in F$, either $e \leq f$ for all $f \in F_0$ or $e \geq f$ for some $f \in F_0$. In the first case, $e \leq \wedge F_0$ and in the second $e \geq f \geq \wedge F_0$. So $F \cup \{\wedge F_0\}$ is totally ordered, i.e. $\wedge F_0 \in F$. $\qquad\qquad\square$

A.3 Topology

Throughout the text, we assume that the reader has a solid knowledge of basic topology, for example as given in [5]. For more advanced topics, we refer to [12] or [21]. In this appendix we give some technical results that are standard, yet not so trivial that they can be used without reference.

Compactness

In a topological space, compactness is defined using open coverings. However, it can also be defined using closed sets. To show this, we first need the following.

Definition A.10 Let X be a topological space and $F \subseteq \mathscr{P}(X)$ a family of subsets. Then F has the **finite intersection property** if for every $\{A_i\}_{i=1}^{n} \subseteq F$ we have that $\bigcap_{i=1}^{n} A_i \neq \emptyset$.

Using this, we can give the equivalent definition of compactness.

Proposition A.11 *Suppose X is a topological space. Then the following are equivalent:*

1. *X is compact.*
2. *Every family $F \subseteq \mathscr{P}(X)$ consisting of closed subsets with the finite intersection property satisfies $\bigcap F \neq \emptyset$.*

We use this in the main text to show that the space of ultrafilters is compact with respect to the ultra topology in Chap. 6.

One of the most important theorems involving compactness is Tychonoff's theorem:

Theorem A.12 (Tychonoff) *Suppose X_i is a non-empty topological space for every $i \in I$. Then $\prod_{i \in I} X_i$ is compact if and only if every X_i is compact.*

The combination of compactness and the Hausdorff property often give strong results, for example in the following lemma.

Lemma A.13 *Suppose X is a compact space and Y is a Hausdorff. Furthermore, let $f : X \to Y$ be a continuous bijection. Then f is a homeomorphism.*

Miscellaneous

Throughout the main text, we also need a few results from topology. The first concerns the separation axiom T_3.

Lemma A.14 *If X is T_3, $U \subseteq X$ is open and $x \in U$, then there is a $V \subseteq X$ open such that $x \in V \subseteq \overline{V} \subseteq U$.*

Next, we have a well-known result about extensions of continuous functions.

Proposition A.15 *Suppose X and Y are topological spaces, where Y is Hausdorff. Furthermore, suppose $A \subseteq X$ is dense and $f, g : X \to Y$ are continuous functions that coincide on A. Then $f = g$.*

Most topological properties are preserved under finite products of topological spaces. However, with infinite products, this is not always the case. However, we do have the following two results, of which the second is the most famous one.

Theorem A.16 *Countable products of metrizable topological spaces are metrizable.*

A.4 Complex Analysis

For an introduction to complex analysis, we refer to [19]. Here, we state a more advanced result: Hurwitz's theorem.

Theorem A.17 (Hurwitz) *Let $G \subseteq \mathbb{C}^m$ be a connected open set and $\{f_n\}_{n \in \mathbb{N}}$ a sequence of holomorphic functions on G that converges uniformly on every compact subset of G to some $f \in H(G)$. Furthermore, suppose that no f_n has zeroes in G. Then either f has no zeroes in G or f is identically zero on G.*

A proof can be found in [14] (Theorem 1.3.8).

Appendix B
Functional Analysis and Operator Algebras

In this appendix we treat a collection of topics from functional analysis and operator algebras that are needed throughout the main text. A more extended survey of these subjects can be found in many texts, for example in [13, 18].

B.1 Basic Functional Analysis

For a normed vector space V, we can consider **bounded linear functionals** on V. These are linear maps $f : V \to \mathbb{C}$ such that

$$\sup_{\|v\|=1} |f(v)| < \infty.$$

We collect all such bounded linear functionals on V in the vector space V^*, which we call the **dual space** of V. This dual space then has a natural norm itself, given by

$$\|f\| = \sup_{\|v\|=1} |f(v)|,$$

for all $f \in V^*$. This gives the dual space a natural topology, but the dual space also has another topology. To describe this topology, we define for all $f \in V^*$, $v \in V$ and $\varepsilon > 0$ the set

$$B(f, v, \varepsilon) = \{g \in V^* \mid |f(v) - g(v)| < \varepsilon\}.$$

It is clear that these sets form a subbase for a topology on V^*, since the union of these sets is clearly all of V^*. We call this topology the **weak*-topology**. One of the most important results about this topology is the following theorem.

Theorem B.1 (Banach-Alaoglu) *Suppose V is a normed vector space. Then the closed unit ball of the dual space V^*, i.e.*

© The Author(s) 2016
M. Stevens, *The Kadison-Singer Property*,
SpringerBriefs in Mathematical Physics 14, DOI 10.1007/978-3-319-47702-2

$$\{f \in V^* \mid \|f\| \leq 1\},$$

is compact with respect to the weak*-topology.

We also have the *Hahn-Banach theorem* for bounded linear functionals, which concerns extensions.

Theorem B.2 (Hahn-Banach) *Suppose V is a normed, complex vector space and W is a linear subspace of V. If $f : W \to \mathbb{C}$ is a bounded functional, then there is an extension $g : V \to \mathbb{C}$ (i.e. $g|_W = f$) such that $\|g\| = \|f\|$.*

The above Hahn-Banach theorem is the one we need in the main text. In fact, there are many theorems that go by the same name. These theorems differ a little in their assumptions, but they all give an extension which preserves some crucial property.

The last fundamental theorem from basic functional analysis that we discuss here concerns convexity. For this, we first need the following definition.

Definition B.3 Suppose V is a vector space and $S \subseteq V$. We define the **convex hull** of S to be:

$$\text{co}(S) = \Big\{ \sum_{i=1}^{n} t_i s_i \ \Big| \ n \in \mathbb{N}, t_i \geq 0, \sum_{i=1}^{n} t_i = 1, s_i \in S \Big\},$$

i.e. the set of all finite convex combinations of elements in K.

Using this definition, we have the following important result.

Theorem B.4 (Krein-Milman) *Suppose V is a normed vector space and $K \subseteq V$ is a convex compact subset. Then:*

$$K = \overline{\text{co}(\partial_e K)}.$$

Furthermore, if $M \subseteq V$ is such that $K = \overline{\text{co}(M)}$, then $\partial_e K \subseteq \overline{M}$.

B.2   Hilbert Spaces

One of the main concepts in the main text is that of a *Hilbert space*.

Definition B.5 A **Hilbert space** H is a complex vector space endowed with a complex inner product $\langle \cdot, \cdot \rangle$, which we take linear in the *second* coordinate, such that H is complete with respect to the norm $\|\cdot\|$ induced by the inner product via $\|x\|^2 = \langle x, x \rangle$.

Hilbert spaces can be seen as generalizations of Euclidean vector spaces. Therefore, we also want to consider bases for Hilbert spaces.

Definition B.6 Suppose H is a Hilbert space. Then a subset $E \subseteq H$ is called a **basis** for H if E is an orthonormal set whose linear span is dense in H.

Note that if the cardinality of a basis of H is finite, then the Hilbert space is isomorphic to a complex Euclidean vector space. We have a special name for Hilbert spaces that have a countable basis.

Definition B.7 H is called **separable** if it has a countable basis.

We also need the notion of *orthogonal families*.

Definition B.8 Let H be a Hilbert space. Two subsets $C, D \subseteq H$ are said to be **orthogonal** if for every $c \in C$ and $d \in D$, $\langle c, d \rangle = 0$. A family of subspaces $\{C_i\}_{i \in I}$ of H is said to be an **orthogonal family** if all pairs of members are orthogonal.

Direct Sums of Hilbert Spaces

Given two Hilbert spaces and H_1 and H_2, we can form a Hilbert space $H = H_1 \oplus H_2$, which has an inner product $\langle \, , \, \rangle$ defined by

$$\langle (x_1, x_2), (y_1, y_2) \rangle = \langle x_1, y_1 \rangle_1 + \langle x_2, y_2 \rangle_2,$$

where $\langle \, , \, \rangle_1$ and $\langle \, , \, \rangle_2$ are the inner products on H_1 and H_2, respectively. H is called the **direct sum** of H_1 and H_2. Conversely, given a Hilbert space H and a closed linear subspace $K \subseteq H$, one can realize H as a direct sum $H = K \oplus K^\perp$, where

$$K^\perp := \{x \in H : \langle x, y \rangle = 0 \ \forall y \in K\}$$

is called the **orthogonal complement** of K.

Operators on Hilbert Spaces

We now want to consider linear operators $T : H \to H'$ between two Hilbert spaces. In fact we are only interested in *bounded* operators.

Definition B.9 Let H be a Hilbert space and $T : H \to H'$ a linear operator. We say that T is **bounded** if there is a $k > 0$ such that $\|T(x)\| \leq k\|x\|$ for all $x \in H$. The set of all bounded operators from H to H' is denoted by $B(H, H')$.

Note that $B(H, H')$ is not just a set, but a normed vector space. Here scalar multiplication and addition are defined pointwise. The norm is naturally given by

$$\|T\| = \sup_{\|x\|=1} \|T(x)\|.$$

Furthermore, for every $T \in B(H, H')$ there is a unique operator $T^* \in B(H', H)$ such that $\langle x, T(y) \rangle = \langle T^*(x), y \rangle$ for every $x \in H'$ and $y \in H$. The operator T^* is called the **adjoint** of T.

When $H = H'$, we write $B(H) := B(H, H)$ and we observe that defining multiplication by composition, i.e. $(TS)(x) = T(S(x))$ for all $T, S \in B(H)$ and $x \in H$, gives $B(H)$ the structure of an algebra.

In the main text we need the following rather technical result.

Proposition B.10 *Suppose that H is a Hilbert space with a basis $\{e_i\}_{i \in I}$. Suppose $a \in B(H)$ and $\alpha > 0$ such that $|\langle x, ay \rangle| \leq \alpha \|x\| \|y\|$ for all $x, y \in H$ with finite support, i.e. for all $x, y \in H$ such that $\{i \in I : \langle x, e_i \rangle \neq 0\}$ and $\{i \in I : \langle y, e_i \rangle \neq 0\}$ are both finite. Then $\|a\| \leq \alpha$.*

Operators on Direct Sums

Note that for a given direct sum $H_1 \oplus H_2$, there are canonical inclusion and projection maps:

$$\iota_1 : H_1 \to H_1 \oplus H_2, \ \iota_1(x) = (x, 0)$$
$$\iota_2 : H_2 \to H_1 \oplus H_2, \ \iota_2(y) = (0, y)$$
$$\pi_1 : H_1 \oplus H_2 \to H_1, \ \pi_1(x, y) = x$$
$$\pi_1 : H_1 \oplus H_2 \to H_2, \ \pi_2(x, y) = y$$

Using this, for given $a_1 \in B(H_1)$ and $a_2 \in B(H_2)$, one can define

$$(a_1, a_2) : H_1 \oplus H_2 \to H_1 \oplus H_2,$$

by $(a_1, a_2) = \iota_1 a_1 \pi_1 + \iota_2 a_2 \pi_2$, i.e. $(a_1, a_2)(x, y) = (a_1(x), a_2(y))$. Clearly, we then have $(a_1, a_2) \in B(H_1 \oplus H_2)$. We can extend this idea for subsets $A_1 \subseteq B(H_1)$ and $A_2 \subseteq B(H_2)$; $A_1 \oplus A_2 \subseteq B(H_1 \oplus H_2)$.

Conversely, one can ask the question whether for some $a \in B(H_1 \oplus H_2)$ there are $a_1 \in B(H_1)$ and $a_2 \in B(H_2)$ such that $a = (a_1, a_2)$. The following proposition answers this question.

Proposition B.11 *Suppose that H_1 and H_2 are Hilbert spaces and $a \in B(H_1 \oplus H_2)$. Then there are $a_1 \in B(H_1)$ and $a_2 \in B(H_2)$ such that $a = (a_1, a_2)$ if and only if $a(\iota_1(H_1)) \subseteq \iota(H_1)$ and $a(\iota_2(H_2)) \subseteq \iota_2(H_2)$.*

In the case that an operator $a \in B(H_1 \oplus H_2)$ can be written as $a = (a_1, a_2)$ for some $a_1 \in B(H_1)$ and $a_2 \in B(H_2)$, we say that a **decomposes over the direct sum** $H_1 \oplus H_2$. Likewise, if an algebra $A \subseteq B(H_1 \oplus H_2)$ satisfies $A = A_1 \oplus A_2$ for some $A_1 \subseteq B(H_1)$ and $A_2 \subseteq B(H_2)$, we say that A decomposes over the direct sum $H_1 \oplus H_2$.

Projection Lattice

Definition B.12 Suppose H is a Hilbert space and $p \in B(H)$. Then p is a **projection** if $p^2 = p^* = p$.

Note that a projection $p \in B(H)$ is always positive, since for any $x \in H$ we have

$$\langle x, px \rangle = \langle x, p^2 x \rangle = \langle x, p^* px \rangle = \langle px, px \rangle = \|px\|^2 \geq 0.$$

Now, if we write $\mathscr{P}(H)$ for the set of all projections in $B(H)$ for a Hilbert space H, it is clear that for any $p \in \mathscr{P}(H)$, we have $1 - p \in \mathscr{P}(H)$. We can now introduce a partial order $\leq$ on $\mathscr{P}(H)$ by saying that $p \leq q$ if and only if $q - p \geq 0$. By the above it follows that (with respect to $\leq$) 0 is the minimal element of $\mathscr{P}(H)$ and 1 is the maximal element. Furthermore, $p \leq q$ is equivalent to $p(H) \subseteq q(H)$.

We need the following technical lemma in the main text. The proof of this is merely a computation.

Lemma B.13 *Suppose p and q are projections on a Hilbert space H such that $p \leq q$. Furthermore, let $x, x' \in H$. Then $\|q(x) - p(x)\| \leq \|q(x) - p(x')\|$.*

In the main text, we are primarily interested in non-zero projections and more specifically in minimal elements of the set of non-zero projections.

Definition B.14 Let H be a Hilbert space and $p \in B(H)$ such that $p \neq 0$. Then p is called a **minimal projection** if $q \in \mathscr{P}(H)$ and $0 \leq q \leq p$ implies $q = 0$ or $q = p$.

B.3 C*-Algebras

We already saw that for a given Hilbert space H the operator algebra $B(H)$ not only has the structure of an algebra, but also has an adjoint operation and a norm. Together, these properties give $B(H)$ a much more special algebraic structure, namely that of a C*-algebra.

Definition B.15 A **C*-algebra** is a normed, associative algebra A endowed with an operation $* : A \to A, a \mapsto a^*$ (we call a^* the **adjoint** of a), with the following compatibility structure:

1. A is **complete** in the norm $\|\cdot\|$.
2. The norm is **submultiplicative**, i.e. $\|ab\| \leq \|a\|\|b\|$ for all $a, b \in A$.
3. The adjoint operation is an **involution**, i.e. $a^{**} = a$ for all $a \in A$.
4. The adjoint operation is **conjugate-linear**, i.e. $(\lambda a + b)^* = \bar{\lambda} a^* + b^*$ for all $\lambda \in \mathbb{C}$ and $a, b \in A$.
5. The adjoint operation is **anti-multiplicative**, i.e. $(ab)^* = b^* a^*$ for all $a, b \in A$.
6. The **C*-identity** holds: $\|a^* a\| = \|a\|^2$ for all $a \in A$.

A C*-algebra A is called **unital** if it contains an algebraic unit 1 (i.e. $a1 = 1a = a$ for all $a \in A$). Since the adjoint is an involution and is anti-multiplicative, automatically $1^* = 1$. By the C*-identity it then also follows that $\|1\| = 1$.

The C*-identity together with submultiplicativity also guarantees a more immediate compatibility between the adjoint operation and the norm.

Lemma B.16 *Suppose A is a C*-algebra. Then the adjoint preserves the norm, i.e. $\|a^*\| = \|a\|$ for all $a \in A$.*

We can also consider C*-subalgebras.

Definition B.17 Let A be a C*-algebra. A **C*-subalgebra** S of A is a subalgebra $S \subseteq A$ that is topologically closed (with the topology coming from the norm $\|\cdot\|$ of A) and closed under the adjoint operation, i.e. $a^* \in S$ for all $a \in S$.

Note that by the conditions on a C*-subalgebra, every C*-subalgebra is a C*-algebra in its own right, by restriction of the norm and adjoint operations to the subalgebra.

Positivity

In the main text we study states. For the definition of states, we need the notion of positive elements of a C*-algebra.

Definition B.18 Suppose A is a C*-algebra, and let $a \in A$. Then we say that a is **positive** if and only if there is a $b \in A$ such that $a = b^*b$. In this case, we write $a \geq 0$.

There are also different ways of defining positivity when A has more structure, as the following lemma shows.

Lemma B.19 *Suppose A is a C*-algebra and let $a \in A$. Then*

- *If A is unital, then a is positive if and only if $a = a^*$ and $\sigma(a) \subseteq [0, \infty)$.*
- *If $A \subseteq B(H)$, then a is positive if and only if $\langle x, ax \rangle \geq 0$ for all $x \in H$.*

Here, $\sigma(a)$ consists of those numbers $\lambda \in \mathbb{C}$ such that $a - \lambda 1$ is not invertible and is called the **spectrum** of a.

The set of positive elements in a C*-algebra A is often denoted by A^+. This set has some special properties.

Proposition B.20 *Suppose A is a C*-algebra. Then:*

- *For any $a \in A$, there are $a_k \geq 0$ such that $a = \sum_{k=0}^{3} i^k a_k$ and $\|a_k\| \leq \|a\|$.*
- *Let $a \in A$ be positive. Then there is a $b \in A^+$ such that $a = b^2$.*
- *Let $a \in A^+$ such that $\|a\| \leq 1$. Then $1 - a^2$ is positive and a commutes with b where $b^2 = 1 - a^2$.*

The notion of positivity also induces a natural partial order $\leq$ on the self-adjoint elements of a C*-algebra A, by defining $b \leq c$ if and only if $0 \leq c - b$. This partial order has the following properties.

Lemma B.21 *If c, d are self-adjoint and $-d \leq c \leq d$, then $\|c\| \leq \|d\|$.*

Lemma B.22 *Suppose H is a Hilbert space and $d \in B(H)$ such that $d \geq 0$ and $\|d\| = 1$, then $d \leq 1$.*

Characters

When considering abelian C*-algebras, characters play a major role.

Definition B.23 Let A be a C*-algebra. A **character** is a non-zero algebra homomorphism $c : A \to \mathbb{C}$, i.e. c is multiplicative and linear. The set of all characters on A is denoted by $\Omega(A)$.

First, we give three properties of characters.

Lemma B.24 *Suppose that A is a unital C*-algebra and $c \in \Omega(A)$. Then:*

- $c(1) = 1$.
- *If $a = a^* \in A$, then $c(a) \in \mathbb{R}$.*
- $c(a^*) = \overline{c(a)}$ *for all $a \in A$.*

Because of the following result, characters are important for abelian C*-algebras.

Theorem B.25 (Gelfand isomorphism) *Suppose that A is a non-zero abelian C*-algebra. Then the map*

$$G : A \to C_0(\Omega(A)), G(a)(f) = f(a),$$

is an isomorphism of C-algebras.*

The following lemma is an easy consequence of the Gelfand isomorphism.

Lemma B.26 *Suppose A is an abelian C*-algebra. Then $\Omega(A)$ separates points.*

One can use this lemma to prove the following result about projections and characters.

Corollary B.27 *Suppose A is a C*-algebra. Then, for every $g \in \Omega(A)$ and projection $p \in A$, $g(p) \in \{0, 1\}$. If $p \in A$ is a non-zero projection, there is a $f \in \Omega(A)$ such that $f(p) = 1$.*

B.4   von Neumann Algebras

In order to define von Neumann algebras, we first introduce the *strong topology*. We do this by means of a subbasis. For every $a \in B(H)$, $x \in H$ and $\varepsilon > 0$, define:

$$S(a, x, \varepsilon) := \{b \in B(H) : \|(a - b)x\| < \varepsilon\}.$$

Collecting these sets together in $\mathscr{S} := \{S(a, x, \varepsilon) : a \in B(H), x \in H, \varepsilon > 0\}$, we obtain a subbasis for a topology on $B(H)$, since $\bigcup \mathscr{S} = B(H)$. We call this topology the **strong topology** on $B(H)$. An important property of this topology is given in terms of convergent nets. See [18] for details.

Proposition B.28 *Let H be a Hilbert space and $\{a_i\}_{i \in I}$ a net in $B(H)$. Furthermore, let $a \in B(H)$. Then the following are equivalent:*

1. $\{a_i\}_{i \in I}$ converges to a with respect to the strong topology on $B(H)$.
2. For each $x \in H$, $\{a_i(x)\}$ converges to $a(x)$.

Using the strong topology, we can directly define von Neumann algebras.

Definition B.29 Let H be a separable Hilbert space. Then a *-subalgebra $A \subseteq B(H)$ is called a **von Neumann algebra** if it is closed with respect to the strong topology.

By now, we have two topologies on $B(H)$; the norm topology and the strong topology. C*-subalgebras deal with the norm topology, whereas von Neumann algebras are defined using the strong topology. The following proposition gives a link between these two different viewpoints.

Proposition B.30 *Let H be a Hilbert space and suppose that $A \subseteq B(H)$ is a von Neumann algebra. Then A is a C*-subalgebra of $B(H)$.*

There is an important result about von Neumann algebras that involves the commutant of an algebra.

Proposition B.31 *Let H be a Hilbert space and $A \subseteq B(H)$ a *-subalgebra. Then A' is a von Neumann algebra.*

In the main text we make use of *generated* von Neumann algebras. For any set $S \subseteq B(H)$ the von Neumann algebra **generated by** S is

$$\langle S \rangle_{vN} := \bigcap \{ A \subseteq B(H) : A \text{ is a von Neumann algebra and } S \subseteq A \},$$

which is in fact a von Neumann algebra since an arbitrary intersection of von Neumann algebras is clearly again a von Neumann algebra.

Projections in von Neumann Algebras

When considering von Neumann algebras, projections play a major role, because of the following proposition.

Proposition B.32 *Suppose H is a separable Hilbert space and $A \subseteq B(H)$ is a von Neumann algebra. Then A is generated by its projections.*

In the main text we need some elementary results about projections and von Neumann algebras, which we state here.

Lemma B.33 *Suppose H is a Hilbert space, $A \subseteq B(H)$ is a von Neumann algebra and $p \in B(H)$ is a projection. Then:*

- $\mathbb{C}p$ *is a von Neumann algebra.*
- *If $p \in A$, then pAp is a von Neumann algebra.*

Appendix C
Additional Material

In this appendix, we use definitions and results from the main text to provide some additional background. These are not included in the main text itself, since they would merely disturb the natural storyline.

C.1 Transitivity Theorem

The following theorem was proven by Kadison [7].

Theorem C.1 (Transitivity theorem) *Suppose A is a non-zero C^*-algebra, acting irreducibly on a Hilbert space H. Furthermore, let $n \in \mathbb{N}$, let $\{x_i\}_{i=1}^n \subseteq H$ be a linearly independent set and let $\{y_i\}_{i=1}^n \subseteq H$ be any subset. Then there exists an $a \in A$ such that $a(x_i) = y_i$ for all $i \in \{1, \ldots, n\}$.*

Furthermore, if there is a $v = v^* \in B(H)$ such that $v(x_i) = y_i$ for every $i \in \{1, \ldots, n\}$, then there is also a $b = b^* \in A$ such that $b(x_i) = y_i$ for all $i \in \{1, \ldots, n\}$.

C.2 G-Sets, M-Sets and L-Sets

As the start of a series of technical results, we begin by defining some important sets associated with states.

Definition C.2 Suppose A is a unital C^*-algebra and $f \in S(A)$. Then define the following subsets of A:

$$N_f = \{a \in A : f(a) = 0\},$$

$$L_f = \{a \in A : f(a^*a) = 0\},$$

© The Author(s) 2016
M. Stevens, *The Kadison-Singer Property*,
SpringerBriefs in Mathematical Physics 14, DOI 10.1007/978-3-319-47702-2

$$G_f = \{a \in A : |f(a)| = \|a\| = 1\},$$

$$M_f = \{a \in A : f(ab) = f(ba) = f(a)f(b) \, \forall b \in A\}.$$

These sets are called the **null-space**, **L-set**, **G-set** and **M-set** of f, respectively. We write G_f^+ for the set of positive elements in G_f.

For a state f, we are especially interested in the structure of the set G_f. To determine this, we use the sets N_f, L_f and M_f. Namely, we have the following sequence of results. For the (straightforward) proofs, see [1].

Lemma C.3 *Suppose A is a unital C^*-algebra, $f \in S(A)$ and $a \in A$. Then:*

1. $M_f \subseteq A$ is a subalgebra.
2. $a \in M_f$ if and only if $a - f(a)1 \in L_f \cap L_f^*$.
3. $G_f \subseteq M_f$.
4. G_f is a semigroup.

For a pure state, there is a nice description of the null-space in terms of the L-set. To give this description, we first give two more properties of states. For more details, see [13].

Lemma C.4 *Suppose A is a C^*-algebra and let $f \in S(A)$. Suppose $a, b \in A$. Then we have the following two properties:*

- $f(a^*a) = 0$ if and only if $f(ba) = 0$ for all $b \in A$.
- $f(b^*a^*ab) \leq \|a^*a\| f(b^*b)$.

We can apply these above properties to describe the algebraic structure of L-sets.

Lemma C.5 *Suppose A is a C^*-algebra and $f \in S(A)$. Then L_f is a left-ideal.*

Proof It is clear that L_f is closed under scalar multiplication. To see that it is closed under addition, suppose $a, b \in L_f$. Then by the Cauchy-Schwarz inequality (Lemma 3.3), we have $f(a^*b) = 0$ and $f(b^*a) = 0$. Therefore,

$$f((a + b)^*(a + b)) = f(a^*a) + f(a^*b) + f(b^*a) + f(b^*b) = 0,$$

i.e. $a + b \in L_f$. Now, again suppose that $a \in L_f$ and let $c \in A$ be arbitrary. Then, applying Lemma C.4,

$$f((ca)^*ca) = f(a^*c^*ca) \leq \|c^*c\| f(a^*a) = 0,$$

so $ca \in L_f$. Hence L_f is a left-ideal. $\qquad\square$

Now, we can make the connection between the notions of null-spaces and L-sets in the case of pure states.

Lemma C.6 *Suppose A is a C^*-algebra and $f \in \partial_e S(A)$. Then $N_f = L_f + L_f^*$.*

Proof First, suppose $a \in L_f$. Then, by the Cauchy-Schwarz inequality,

$$|f(a)|^2 = |f(1^*a)|^2 \le f(1^*1)f(a^*a) = 0,$$

so $a \in N_f$, i.e. $L_f \subseteq N_f$. Likewise, $L_f^* \subseteq N_f$, so by linearity of f, $L_f \subseteq L_f^* \subseteq N_f$.

To show that $N_f \subseteq L_f + L_f^*$, we use the GNS-representation for f, as discussed in Sect. C.3. First of all, since f is pure, the space A/L_f is a Hilbert space with respect to the inner product $(a + L_f, b + L_f) = f(a^*b)$. Furthermore, since f is pure, the map

$$\varphi_f : A \to B(A/L_f), \varphi_f(a)(b + L_f) = ab + L_f$$

has the property that $\varphi_f(A)$ acts irreducibly on A/L_f, by Proposition C.8.

Now suppose $a \in N_f$ is self-adjoint. Then we have

$$(1 + L_f, a + L_f) = f(1^*a) = f(a) = 0,$$

i.e. $1 + L_f$ and $a + L_f$ are linearly independent. Therefore, by the transitivity Theorem C.1, there is a self-adjoint element $v \in \varphi_f(A)$ such that $v(a + L_f) = a + L_f$ and $v(1 + L_f) = 0$. Then $v = \varphi_f(b)$ for some $b \in A$. Define $c = \frac{b^*+b}{2}$. Then $c = c^*$ and

$$\varphi_f(c) = \varphi_f\left(\frac{b^*+b}{2}\right) = \frac{\varphi_f(b)^* + \varphi_f(b)}{2} = \frac{v^*+v}{2} = v,$$

so we have

$$ca + L_f = \varphi_f(c)(a + L_f) = v(a + L_f) = a + L_f,$$

and

$$c + L_f = \varphi_f(c)(1 + L_f) = v(1 + L_f) = 0,$$

i.e. $ca - a \in L_f$ and $c \in L_f$. Define $d := ca - a \in L_f$. Then since $a = a^*$,

$$a = ca - d = (ca - d)^* = ac - d^*.$$

Since $c \in L_f$ and L_f is a left-ideal by Lemma C.5, $ac \in L_f$. Furthermore, $-d^* \in L_f^*$, so $a = ac - d^* \in L_f + L_f^*$.

So, if we take an arbitrary $x \in N_f$, we have $x = x_1 + ix_2$, with $x_1 = \frac{x+x^*}{2} \in N_f$ and $x_2 \in \frac{x-x^*}{2i} \in N_f$. Hence, by the above, $x_1 = y_1 + w_1^*$ and $x_2 = y_2 + w_2^*$ for some $y_1, w_1, y_2, w_2 \in L_f$. Then, $y_1 + y_2 \in L_f$ and $-i(w_1 + w_2) \in L_f$, so

$$x = y_1 + y_2 + (-i(w_1 + w_2))^* \in L_f + L_f^*.$$

Therefore, $N_f \subseteq L_f + L_f^*$, i.e. $N_f = L_f + L_f^*$, as desired. $\qquad\square$

Of course, we are going to apply the above discussion to extensions of pure states, in order to say something about the classification of subalgebras that satisfy the Kadison-Singer property. Therefore, the following result is useful, which states that L- and M-sets behave nicely with respect to extensions.

Lemma C.7 *Suppose H is a Hilbert space and $A \subseteq B(H)$ is a C^*-subalgebra. Furthermore, suppose $g \in \mathrm{Ext}(f)$. Then $L_f \subseteq L_g$ and $M_f \subseteq M_g$.*

Proof Suppose $a \in L_f$. Then $a \in A \subseteq B(H)$ and $f(a^*a) = 0$. Since $g \in \mathrm{Ext}(f)$, and $a^*a \in A$, $g(a^*a) = f(a^*a) = 0$, i.e. $a \in L_g$ and $L_f \subseteq L_g$.

Now suppose $a \in M_f$. Then $a - f(a)1 \in L_f \cap L_f^*$, by Lemma 2. Since we have assumed that $g \in \mathrm{Ext}(f)$, $g(a) = f(a)$, and by the above, $L_f \cap L_f^* \subseteq L_g \cap L_g^*$. Therefore, $a - g(a)1 \in L_g \cap L_g^*$ and hence $a \in M_g$, again by Lemma 2. Hence $M_f \subseteq M_g$, as desired.   □

C.3 GNS-Representation

Next, we treat the so-called *Gelfand-Naimark-Segal representation*. For this, we fix a certain C^*-algebra A and we let $f : A \to \mathbb{C}$ be a state. In Definition C.2, we defined the L-set of f to be

$$L_f = \{a \in A : f(a^*a) = 0\},$$

and in Lemma C.5 we showed that L_f is a left ideal of A.

Now, we note that we have a well-defined inner product on A/L_f, given by

$$\langle a + L_f, b + L_f \rangle = f(a^*b).$$

We can then complete A/L_f to a Hilbert space H_f. Then, we define a map

$$\psi_f : A \times A/L_f \to A/L_f,$$

by setting $\psi_f(a, b + L_f) = ab + L_f$. Since A/L_f is dense in H_f and $\psi_f(a, \cdot)$ is bounded for every $a \in A$, ψ_f uniquely extends to a map $\psi_f' : A \times H_f \to H_f$. Then, we have the map

$$\varphi_f : A \to B(H_f),$$

defined by $\varphi_f(a)(x) = \psi_f'(a, x)$. In fact, φ_f is a *-homomorphism, and as such, it is a representation, which we call the **Gelfand-Naimark-Segal representation** belonging to f.

The main result we use about the GNS-representation is the following:

Proposition C.8 *Suppose A is a C^*-algebra and $f \in S(A)$. Then $f \in \partial_e S(A)$ if and only if $\varphi_f(A)$ acts irreducibly on H_f.*

C.4   Miscellaneous

In Sects. C.1 and C.3, we discussed some fundamental results, which are treated in many texts on operator algebras. In this section, we give results which are less well-known.

State-Like Functionals

As we already mentioned in Sect. B.1, there are many theorems similar to the Hahn-Banach theorem. There is also a theorem for C*-algebras in which 'positivity' is preserved. For this, we need the notion of *state-like functionals*.

Definition C.9 Suppose A is a unital C*-algebra and $C \subseteq A$ is a self-adjoint linear subspace of A that contains the unit. Then a linear map $f : C \to \mathbb{C}$ that satisfies $f(c^*) = \overline{f(c)}$ for every $c \in C$, $f(c) \geq 0$ for every positive $c \in C$ and $f(1) = 1$, is called a **state-like functional** on C. The set of all state-like functionals on C is written as $\mathrm{SLF}(C)$.

For these state-like functionals, we have the following extension theorem, which resembles the Hahn-Banach theorem. For its proof, we refer to ([4, 2.10.1]).

Theorem C.10 *Suppose A is a unital C*-algebra and $C \subseteq A$ is a self-adjoint linear subspace that contains the unit. Suppose $f : C \to \mathbb{C}$ is a state-like functional. Then there is a state-like functional $g : A \to \mathbb{C}$ that extends f.*

The Projection Lattice in the Strong Topology

In Sect. B.2, we discussed some properties of the projection lattice for a Hilbert space. In Sect. B.4 we saw that projections play a major role for von Neumann algebras. Since von Neumann algebras are defined using the strong topology, we need some result about the projection lattice with respect to the strong topology. Here, for a Hilbert space H and a subset $Y \subseteq B(H)$, $\mathrm{Cl}_{\mathrm{str}}(Y)$ denotes the strong closure of Y.

We first have the following result.

Proposition C.11 *Suppose F is a totally ordered family of projections on a Hilbert space H. Then $\vee F \in \mathrm{Cl}_{\mathrm{str}}(F)$.*

Proof Write $\lambda = \vee F$ and consider $A = \bigcup_{p \in F} p(H)$.

For $a, b \in A$ there are $p, q \in F$ such that $a \in p(H)$ and $b \in q(H)$. Since F is totally ordered, we can assume without loss of generality that $p \leq q$. Then we have that $a \in p(H) \subseteq q(H)$, so $a, b \in q(H)$, whence $a + b \in q(H) \subseteq A$. Furthermore, for $\mu \in \mathbb{C}$ and $a \in A$, there is a $p \in F$ such that $a \in p(H)$, whence $\mu a \in p(H) \subseteq A$. Therefore, A is a linear subspace of H.

Hence $\overline{A}$ is a closed linear subspace of H. We now claim that $\lambda(H) = \overline{A}$.

First, let q be the projection onto $\overline{A}$. Then for all $p \in F$, $p(H) \subseteq A \subseteq \overline{A} = q(H)$, i.e. $p \leq q$ for all $p \in F$. Therefore, $\lambda \leq q$, so $\lambda(H) \subseteq q(H) = \overline{A}$. For the converse, observe that for any $p \in F$, we have $p \leq \lambda$, i.e. $p(H) \subseteq \lambda(H)$, so we obtain that $A = \bigcup_{p \in F} p(H) \subseteq \lambda(H)$. Therefore, $\overline{A} \subseteq \overline{\lambda(H)} = \lambda(H)$. So, indeed, $\lambda(H) = \overline{A}$.

Now, let $x \in H$. Then $\lambda(x) \in \overline{A}$, so there is a sequence $\{y_{x,n}\}_{n=1}^{\infty} \subseteq A$ such that $\lim_{n \to \infty} y_{x,n} = \lambda(x)$. For all $n \in \mathbb{N}$ there is a $p_{x,n} \in F$ such that we have $y_{x,n} = p(z_{x,n})$ for some $z_{x,n} \in H$. So, for every $\varepsilon > 0$, there is a $n_\varepsilon \in \mathbb{N}$ such that $\|\lambda(x) - p_{x,n_\varepsilon}(z_{x,n_\varepsilon})\| < \varepsilon$.

By Lemma B.13 we conclude that

$$\|\lambda(x) - p_{x,n_\varepsilon}(x)\| \le \|\lambda(x) - p_{x,n_\varepsilon}(z_{x,n_\varepsilon})\| < \varepsilon.$$

Now, for any $q \ge p_{x,n_\varepsilon}$, we have that $\lambda - q \le \lambda - p_{x,n_\varepsilon}$, so

$$\|\lambda(x) - q(x)\| \le \|\lambda(x) - p_{x,n_\varepsilon}(x)\| < \varepsilon.$$

Since $\varepsilon > 0$ was arbitrary, $\lambda(x) = \lim_{p \in F} p(x)$. Since $x \in H$ was arbitrary, we therefore conclude that λ is the strong limit of the net $\{p\}_{p \in F} \subseteq F$, i.e. we have $\lambda \in \mathrm{Cl}_{\mathrm{str}}(F)$.   □

For the next result on the projection lattice and the strong topology, we first need the following lemma.

Lemma C.12 *Suppose H is a Hilbert space and let $F \subseteq \mathscr{P}(H)$ be some family of projections. Then we have*

$$\bigvee_{p \in F} \{1 - p\} = 1 - \bigwedge_{p \in F} \{p\}.$$

Proof For all q, $q \ge \wedge\{p\}$, so $1 - q \le 1 - \wedge\{p\}$, whence $\vee\{1 - p\} \le 1 - \wedge\{p\}$.

For all q, $\vee\{1 - p\} \ge 1 - q$, so $1 - \vee\{1 - p\} \le q$, whence $1 - \vee\{1 - p\} \le \wedge\{p\}$, i.e. $\vee\{1 - p\} \ge 1 - \wedge\{p\}$.

Therefore, $\vee\{1 - p\} = 1 - \wedge\{p\}$.   □

Using this lemma, we can prove the following.

Corollary C.13 *Suppose F is a totally ordered family of projections on a Hilbert space H. Then $\wedge F \in \mathrm{Cl}_{\mathrm{str}}(F)$.*

Proof Consider the family $G := \{1 - p : p \in F\}$, which is again a totally ordered family of projections on H. By Proposition C.11, then $\vee G \in \mathrm{Cl}_{\mathrm{str}}(G)$. By Lemma C.12, $\vee G = 1 - \wedge F$, i.e. $1 - \wedge F \in \mathrm{Cl}_{\mathrm{str}}(G)$. Therefore, there is a net $\{g_i\}_{i \in I} \subseteq G$ such that $1 - \wedge F$ is the strong limit of $\{g_i\}_{i \in I}$. However, for every $i \in I$, $g_i = 1 - p_i$ for a certain $p_i \in F$.

Now suppose $x \in H$ and $\varepsilon > 0$. Then there is a $i_0 \in I$ such that for every $i \ge i_0$,

$$\|((1 - \wedge F) - g_i)(x)\| < \varepsilon.$$

Then we also obtain for every $i \geq i_0$ that

$$\|(p_i - \wedge F)(x)\| = \|((1 - \wedge F) - (1 - p_i))(x)\| = \|((1 - \wedge F) - g_i)(x)\| < \varepsilon.$$

Therefore, $\lim_{i \in I} p_i(x) = \wedge F(x)$, i.e. $\wedge F$ is the strong limit of the net $\{p_i\}_{i \in I}$ in F, so $\wedge F \in \text{Cl}_{\text{str}}(F)$. $\qquad\square$

Appendix D
Notes and Remarks

In this appendix, we comment on the things we discussed in the main text. First, we give some very specific notes about technicalities in the main text. Subsequently we will make some remarks which have a broader context.

Notes per Chapter

Chapter 2

In Chap. 2 we defined states for both the algebra M and the subalgebra D. Of course, these definitions are alike and can be generalized. This is done in Chap. 3.

The unique extension given in the proof of Theorem 2.14, is in fact given by a conditional expectation. We discuss these conditional expectations in more detail in Sect. D, but the idea is the following: consider the map diag : $M \to D$, given by $\mathrm{diag}(a)_{ii} = \langle e_i, ae_i \rangle$. This map is linear, unital and satisfies diag $\circ\, i\, =$ Id, where $i : D \hookrightarrow M$ is the inclusion map. Then the unique extension of pure states is given by the pullback of the map diag, i.e. for a pure state $f \in \partial_e S(D)$, $g := f \circ \mathrm{diag}$ is the unique pure extension.

Chapter 4

The result of Proposition 4.4 is to be expected when carefully using Lemma 4.3. Namely, for some Hilbert space H and $A_1, A_2 \in C(B(H))$, such that $A_1 \subseteq A_2$, Lemma 4.3 gives

$$A_1 \subseteq A_2 \subseteq A_2' \subseteq A_1',$$

so in fact we have

$$A_2' \setminus A_2 \subseteq A_2' \setminus A_1 \subseteq A_1' \setminus A_1,$$

i.e. if we define the map $\varphi : C(B(H)) \to \mathscr{P}(B(H))$ by $\varphi(A) = A' \setminus A$, we see that φ is an anti-homomorphism between the partially ordered sets $C(B(H))$ and $\mathscr{P}(B(H))$. Therefore, maximality in the poset $C(B(H))$ corresponds to minimality in the poset $\varphi(C(B(H))) \subseteq \mathscr{P}(B(H))$. In fact, Proposition 4.4 shows exactly that maximality in $C(B(H))$ corresponds to the element $\emptyset \in \varphi(C(B(H)))$, so the minimal element of $\varphi(C(B(H)))$ is the minimal element of $\mathscr{P}(B(H))$.

© The Author(s) 2016
M. Stevens, *The Kadison-Singer Property*,
SpringerBriefs in Mathematical Physics 14, DOI 10.1007/978-3-319-47702-2

In the proof of Theorem 4.5, we show that if $A, C \in C(B(H))$, $A \subseteq C$ and A has the Kadison-Singer property, then necessarily $A = C$. We do this by first showing that $A \cong C$, followed by showing that the inclusion $i : A \hookrightarrow C$ is in fact giving this isomorphism. One might think that $A \subseteq C$ and $A \cong C$ already implies that $A = C$. However, this is not the case. As an example, consider the subalgebra

$$\ell^\infty(2\mathbb{N}) := \{f \in \ell^\infty(\mathbb{N}) \mid f(2n - 1) = 0 \ \forall n \in \mathbb{N}\}.$$

Clearly, $\ell^\infty(2\mathbb{N}) \cong \ell^\infty(\mathbb{N})$, but these two algebras are not the same.

Chapter 5

We restrict ourselves to the case of separable Hilbert spaces. This may seem to be a major restriction, but some remarks can be made justifying this restriction. First of all, in applications of operator algebras within the context of physics (most notably that of quantum theory), non-separable Hilbert spaces almost never play a role. Furthermore, the ungraspability of the non-separable case is a big mathematical issue. After all, we are restricting ourselves to the separable case, since we can make a classification of maximal abelian subalgebras of $B(H)$ where H is a separable Hilbert space (Corollary 5.25). For the non-separable case(s), such a classification is not available so far.

The ideas behind the classification in the separable case (Corollary 5.25) are exactly those of Kadison and Ringrose ([16, 9.4.1]). We expanded and clarified some of their technical arguments.

Chapter 6

The Stone-Čech compactification of a Tychonoff-space can in fact also be constructed using the theory of operator algebras. In fact, for such a space X, its Stone-Čech compactification can be realised as the $\Omega(C_b(X))$, i.e. the character space of the algebra of bounded continuous functions on X. Namely, assuming that the Stone-Čech compactification βX exists for some topological space X, we can show that $C_b(X) \cong C(\beta X) = C_0(\beta X) = C_b(\beta X)$ in the following way. Suppose that $f \in C_b(X)$ and let $D := \{z \in \mathbb{C} \mid |z| \leq \|f\|\}$. Then D is a compact Hausdorff space, and $f : X \to D$ is a continuous function. Therefore, by the universal property of the Stone-Čech compactification, there is a unique continuous $\beta f : \beta X \to D$ that extends f. Hence, we get a well-defined map $\Phi : C_b(X) \to C(\beta X)$, $f \mapsto \beta f$. Since any continuous function on the compact Hausdorff space βX is automatically bounded, we also get a map $\Psi : C(\beta X) \to C_b(X)$, $h \mapsto h|_X$. By the universal property of the Stone-Čech compactification it is clear that Φ and Ψ are each other's inverse, whence $C_0(\beta X) = C(\beta X) \cong C_b(X)$. However, by the Gelfand-isomorphism, we also have $C_b(X) \cong C_0(\Omega(C_b(X)))$, so $\beta X \cong \Omega(C_b(X))$.

Chapter 7

The whole point of introducing and using the Stone-Čech compactification is in the proof of Theorem 7.10. The switch from $\mathbb{N}$ to $\mathrm{Ultra}(\mathbb{N})$, i.e. from a non-compact space to a compact space, exactly gives us that $\partial_e S(A_c)$ is already contained in

$(\beta H')(\mathrm{Ultra}(\mathbb{N}))$, instead of $\overline{(\beta H')(\mathrm{Ultra}(\mathbb{N}))}$. The latter space is bigger and we cannot describe it properly. However, for $(\beta H')(\mathrm{Ultra}(\mathbb{N}))$ itself, we have results like Proposition 7.11.

The Use of Existing Literature

This thesis has one goal: proving Corollary 8.35. Every part of the text is necessary for reaching this goal and we have tried to keep the text as self-contained as possible. The text can be divided in a few parts, each with their own character, their own (intermediate) goal and their own roots in existing literature.

First of all, the introductory Chaps. 2 and 3 together form the foundation for the thesis and have the goal of introducing the necessary concepts for the final classification. The idea of the question can mainly be found in the original article by Kadison and Singer [8], although they spoke of unique pure state extensions instead of the Kadison-Singer property, like we do. In fact, this way of defining the Kadison-Singer property as a property of an algebra is something we added to the theory.

The second part (Chap. 4) contains the first reduction step: maximality is necessary for the Kadison-Singer property. This is also already in [8]. However, we give our own proof of this fact.

Subsequently, Chap. 5 reduces the classification even further, using the classification of maximal abelian von Neumann algebras. This classification is based on an idea of John von Neumann, but there are not many sources for well-written proofs. We have based ourselves on the proof of Kadison and Ringrose in [15, 16]. Although their ideas are exactly those that are behind our proof, we have expanded the proof, by making clear distinctions between the several cases.

Chapters 6 and 7 together reduce the classification to the Kadison-Singer conjecture. The theory of ultrafilters and the Stone-Čech compactification of discrete spaces can be found in many textbooks on topology, but our extensive study of the Stone-Čech compactification for arbitrary Tychonoff spaces has its roots in [21]. The results in Chap. 7 also have one clear source: the article by Joel Anderson [1]. Although this article already gives a much clearer proof of the fact that the continuous subalgebra does not have the Kadison-Singer property than Kadison and Singer do in their article (viz. [8]), we have clarified this even further. Our main improvement concerns the distinction between using the universal property of the Stone-Čech compactification for the Haar states and using the same property, but then for the restricted Haar states. Furthermore, we have not proven all results that in fact hold for arbitrary algebras, but have restricted ourselves to the continuous subalgebra, which gives easier proofs in Sect. 7.3.

In Chap. 8 we complete the classification. For this, we use an article of Tao [20]. He has already simplified the works of Marcus et al. [11], whence we have not concretely used their articles. However, the article contained a minor mistake in the proof of Lemma 8.20. After a short correspondence, Terence Tao improved his argument. Subsequently, we have made an even further simplification for this proof.

Broader Remarks

The Anderson Operator

Throughout the main text we used several technical arguments. Most of them were to be expected within their context. However, in Chap. 7, we used the Stone-Čech compactification of $\mathbb{N}$, which is a *discrete* space, to say things about the *continuous* subalgebra. At first sight, this seems paradoxical, but it is not really. After all, we use $\mathbb{N}$ in order to enumerate the Haar functions. Therefore, it is not the discreteness of $\mathbb{N}$ that is important, but its cardinality, since the continuous subalgebra acts on the separable Hilbert space $L^2(0, 1)$.

Therefore, one might think the same arguments are applicable to the discrete subalgebra. In fact, a lot of structure described in Chap. 7 can be transfered to the case of the discrete subalgebra. This can best be described by means of the following diagram:

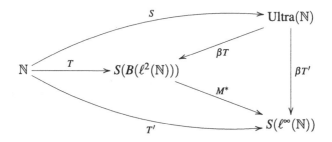

Here, T' and T are defined by $T'(n)(f) = f(n)$ and $T(n)(a) = \langle \delta_n, a\delta_n \rangle$. Furthermore, $\beta T'$ and βT are the continuous extensions of T' and T respectively, obtained by the universal property of the Stone-Čech compactification. Like in Chap. 7, S is the map that assigns the principal filter to every natural number, i.e. $S(n) = F_n$. Lastly, M^* is the pullback of the multiplication operator $M : \ell^\infty(\mathbb{N}) \hookrightarrow B(\ell^2(\mathbb{N}))$.

This diagram is similar to the situation we had in Chap. 7, where the role of T was taken by H and the role of T' by H'. Now, again, $M^* \circ T = T'$ and therefore $M^* \circ \beta T = \beta T'$. Therefore, the above diagram is commutative.

It is easy to see that $T'(\mathbb{N})$ is a total set of states, whence $(\beta T')(\mathrm{Ultra}(\mathbb{N}))$ is a total set of states. Therefore, $\partial_e S(\ell^\infty(\mathbb{N})) \subseteq (\beta T')(\mathrm{Ultra}(\mathbb{N}))$. However, to conclude things about the uniqueness of pure state extensions, we need some kind of injectivity of M^*. However, the above diagram gives no further information, since the set $T(\mathbb{N})$ is not a total set of states: there are operators $a \in B(\ell^2(\mathbb{N}))$ which have a positive diagonal part but are not positive themselves. Therefore, we cannot conclude that all pure states on $B(\ell^2(\mathbb{N}))$ lie in the image of βT.

The high point of Chap. 7 was reached when we defined the Anderson operator. This operator was defined using a bijection $\varphi : \mathbb{N} \to \mathbb{N}$ that had no fixed points. In fact, the bijection that was used respected the structure of the basis formed by the Haar functions, since it permutes groups of Haar functions whose supports are of equal length.

In the case of the discrete subalgebra, we can again consider some bijection $\varphi : \mathbb{N} \to \mathbb{N}$ without fixed points and use this to construct an operator V_φ like the Anderson-operator: we set $V_\varphi(\delta_n) = \delta_{\varphi(n)}$ and extend this linearly to all of $\ell^2(\mathbb{N})$. Then V_φ is unitary, since it permutes a basis and for all $n \in \mathbb{N}$ we have $T(n)(V_\varphi) = 0$.

However, for any $m \in \mathbb{N}$, we have $\|M_{\delta_m} V_\varphi M_{\delta_m}\| = 0$, since φ has no fixed points. This is in contrast to Proposition 7.18. We note here that we have taken δ_m as a projection, which is in fact a minimal projection. This observation becomes particularly interesting when also noting that the main difference between the continuous and the discrete subalgebras is the existence of minimal projections: the continuous subalgebra has none, whereas the discrete subalgebra is even generated by its minimal projections, as we showed in Chap. 5. In fact, for any choice of φ above, there is a non-minimal projection $p \in \ell^\infty(\mathbb{N})$ such that $\|M_p V_\varphi M_p\| = 1$.

Therefore, we are led to believe that the technique of using the Anderson operator in Chap. 7 works precisely since the continuous subalgebra has no minimal projections.

Normal States

In Chap. 2, we described all states on the matrix algebra $M_n(\mathbb{C})$ using density operators. In fact, using the spectral decomposition of density operators, we saw that every state on $M_n(\mathbb{N})$ was given by

$$\omega(a) = \sum_{i=1}^{n} p_i \langle v_i, a v_i \rangle,$$

where $\{v_i\}_{i=1}^{n}$ is some orthonormal basis of $\mathbb{C}^n$ and $\{p_i\}_{i=1}^{n} \subseteq [0, 1]$ is such that $\sum_{i=1}^{n} p_i = 1$. We can generalize these states to the infinite dimensional case. For any orthonormal base $\{v_i\}_{i=1}^{\infty}$ of $\ell^2(\mathbb{N})$ and any sequence $\{p_i\}_{i=1}^{\infty} \subseteq [0, 1]$ such that we have $\sum_{i=1}^{\infty} p_i = 1$, the functional $f : B(\ell^2(\mathbb{N})) \to \mathbb{C}$ defined by

$$f(a) = \sum_{i=1}^{\infty} p_i \langle v_i, a v_i \rangle,$$

is a state on $B(\ell^2(\mathbb{N}))$. Such states are called *normal states* (see [9]). In contrary to the finite dimensional case, the set of normal states do not exhaust the set of all states on $B(\ell^2(\mathbb{N}))$.

It is clear that for any orthogonal set of projections $\{e_i\}_{i \in I}$ we have

$$f\left(\bigvee_{i \in I} e_i\right) = \sum_{i \in I} f(e_i)$$

for a normal state f. In contrast to this, *singular states* are states that annihilate all one-dimensional projections, and thereby all compact operators.

An arbitrary state on $B(\ell^2(\mathbb{N}))$ can be written as a convex combination of a normal and a singular state (as a consequence of Theorem 10.1.15(iii) in [16]). This has an interesting consequence for the concept of pure state extensions. Namely, suppose $n \in \mathbb{N}$ and let $f_n : \ell^\infty(\mathbb{N}) \to \mathbb{C}$ be given by $f_n(a) = a(n)$. Then certainly, $f_n \in \Omega(\ell^\infty(\mathbb{N})) = \partial_e S(\ell^\infty(\mathbb{N}))$. Then, suppose $g \in \text{Ext}(f_n)$ is a pure state. g can be written as a convex combination of a normal and a singular state, but it is pure, so it is either normal or singular. Since g is an extension of f_n, g is non-zero on the projection onto the span of δ_n, so g is not singular. Hence it is normal. So $g(a) = \sum_{i=1}^\infty p_i \langle v_i, a v_i \rangle$ for some orthonormal basis $\{v_i\}_{i=1}^\infty$ and sequence $\{p_i\}_{i=1}^\infty$ such that $\sum_{i=1}^\infty p_i$. Similar to the finite dimensional case, the fact that g is pure then implies that there must be some $i \in \mathbb{N}$ such that $p_i = 1$ and $p_j = 0$ for all $j \neq i$. Therefore, $g(a) = \langle v_i, a v_i \rangle$. However, since $g \in \text{Ext}(f_n)$, we then get $|\langle v_i, \delta_n \rangle| = 1$, whence $g = g_n$, where $g_n(a) := \langle \delta_n, a \delta_n \rangle$ for all $a \in B(\ell^2(\mathbb{N}))$. Therefore, f_n has a unique pure state extension and since $\partial_e \text{Ext}(f_n) = \text{Ext}(f_n) \cap \partial_e S(B(\ell^2(\mathbb{N})))$ by Lemma 3.13, we know that $\partial_e \text{Ext}(f_n) = \{g_n\}$. Since $\text{Ext}(f_n)$ is a closed subset of $S(B(\ell^2(\mathbb{N})))$ it is a compact and convex set, so by the Krein-Milman theorem, $\text{Ext}(f_n) = \{g_n\}$.

Conditional Expectations

In the finite dimensional case (Theorem 2.14) we saw that the unique extension of a pure state is given by its pullback under the map which takes its diagonal part. In fact, for the infinite dimensional case, the same result holds (see Corollary 8.33). Here, we generalize this concept to so-called *conditional expectations*.

For a Hilbert space H and an abelian subalgebra $A \subseteq B(H)$ we say that a map $d : B(H) \to A$ is a *conditional expectation* for A if it is linear, positive and satisfies $d \circ i = \text{Id}$, where $i : A \hookrightarrow B(H)$ is the inclusion.

For a conditional expectation d for A and a state $f \in S(A)$, it is then clear that $f \circ d \in \text{Ext}(f)$. Formulated differently, the pullback $d^* : S(A) \to S(B(H))$, defined by $d^*(f) = f \circ d$, can be considered an extension map.

Therefore, it is natural to ask whether two different conditional expectations give different extension maps. More precisely, suppose d_1 and d_2 are both conditional expectations for A and suppose A has the Kadison-Singer property. Then we have that $f \circ d_1 = f \circ d_2$ for all $f \in \partial_e S(A)$, so $f(d_1(b)) = f(d_2(b))$ for all $b \in B(H)$ and for all $f \in \partial_e S(A)$. However, $\partial_e S(A) = \Omega(A)$ separates points, so $d_1(b) = d_2(b)$ for all $b \in B(H)$. Therefore, $d_1 = d_2$. So, if A has the Kadison-Singer property, then it has at most one conditional expectation.

In fact, Anderson showed ([1, Theorem 3.4]) that if A has the Kadison-Singer property, then A has a conditional expectation. Therefore, if A has the Kadison-Singer property, then it has precisely one conditional expectation.

In the original article by Kadison and Singer ([8, Theorem 2]), it is shown that the continuous subalgebra has more than one conditional expectation. This is proven using very technical arguments, which we find not insightful. The article of Anderson [1] is more helpful and serves as the base for Chap. 7 of this text.

Although we proved that the discrete subalgebra has the Kadison-Singer property in Chap. 8 and that this implies that $\ell^\infty(\mathbb{N})$ has a unique conditional expectation, we can also prove the latter directly. It is implied by the fact that every point-evaluation

$$f_n \in \partial_e S(\ell^\infty(\mathbb{N})), \; f_n(a) = a(n)$$

with $n \in \mathbb{N}$ has a unique extension, given by

$$g_n \in \partial_e S(B(\ell^2(\mathbb{N}))), \; g_n(a) = \langle \delta_n, a\delta_n \rangle.$$

Namely, if d is a conditional expectation for $\ell^\infty(\mathbb{N})$, then for any $a \in B(\ell^2(\mathbb{N}))$ and $n \in \mathbb{N}$ we have

$$d(a)(n) = f_n(d(a)) = (f_n \circ d)(a) = g_n(a) = \langle \delta_n, a\delta_n \rangle,$$

and defining the map d by $d(a)(n) = \langle \delta_n, a\delta_n \rangle$ in fact defines a conditional expectation. Therefore, $\ell^\infty(\mathbb{N})$ has a unique conditional expectation.

Acknowledgments Lastly, I would like to thank the people who have helped and supported me in writing this book and the master thesis that this book is based on.

First of all and most importantly, I would like to thank Klaas Landsman for supervising my master thesis project, for advising me in all aspects of the writing process, for writing the foreword in this book and for supporting me during this whole project.

Secondly, I am grateful to Michael Müger for helping me with some mathematical technicalities when I wrote my master thesis and for being the second reader of the aforementioned thesis.

Last but not least, I thank Serge Horbach for proofreading the first few chapters and for supporting me as a friend.

References

1. Anderson, J.: Extensions, restrictions and representations of states on C*-algebras. Trans. Am. Math. Soc. **249**(2), 303–329 (1979)
2. Anton, H.: Elementary Linear Algebra with Supplemental Applications, 11th edn. Wiley, Hoboken (2014)
3. Davey, B., Priestley, H.: Introduction to Lattices and Order, 1st edn. Cambridge University Press (1990)
4. Dixmier, J.: C*-Algebras. North-Holland Publishing (1977)
5. Gamelin, T., Greene, R.: Introduction to Topology, 2nd edn. Dover Publications (1999)
6. Jänich, K.: Linear Algebra. Springer (1994)
7. Kadison, R.V.: Irreducible operator algebras. Proc. Natl. Acad. Sci. USA **43**(3), 273–276 (1957)
8. Kadison, R.V., Singer, I.: Extensions of pure states. Am. J. Math. **81**(2), 383–400 (1959)
9. Landsman, N.: Mathematical Topics Between Classical and Quantum Mechanics. Springer Monographs in Mathematics. Springer (1998)
10. Lang, S.: Linear Algebra, 2nd edn. World Student Series. Addison-Wesley, Reading, Mass. (1970)
11. Marcus, A., Spielman, D., Srivastava, N.: Interlacing families II: mixed characteristic polynomials and the Kadison-Singer problem. Ann. Math. **182**, 327–350 (2015)
12. Müger, M.: Topology for the working mathematician. http://www.math.ru.nl/~mueger/topology.pdf (2015)

13. Murphy, G.: C*-Algebras and Operator Theory. Academic Press (1990)
14. Rahman, Q., Schmeisser, G.: Analytic Theory of Polynomials. Oxford University Press (2002)
15. Kadison, R.V., Ringrose, J.R.: Fundamentals of the theory of operator algebras. In: Pure and Applied Mathematics. Elementary Theory, vol. 1. Academic Press (1983)
16. Kadison, R.V., Ringrose, J.R.: Fundamentals of the theory of operator algebras. In: Pure and Applied Mathematics. Advanced Theory, vol. 2. Academic Press (1986)
17. Roman, S.: Advanced Linear Algebra, 3rd edn. Springer (2008)
18. Rudin, W.: Functional Analysis. McGraw-Hill Series in Higher Mathematics. McGraw-Hill (1973)
19. Stein, E.M., Shakarchi, R.: Complex Analysis. Princeton University Press (2003)
20. Tao, T.: Real stable polynomials and the Kadison-Singer problem. https://terrytao.wordpress.com/2013/11/04/real-stable-polynomials-and-the-kadison-singer-problem/ (2013)
21. Willard, S.: General Topology. Dover (2004)

Printed in the United States
By Bookmasters